Nuclear Energy

PUES-19

Pergamon Unified
Engineering Series

Pergamon
Unified Engineering
Series

Nuclear Energy

An Introduction to the Concepts, Systems, and Applications of Nuclear Processes

---

Raymond L. Murray
Nuclear Engineering Department,
North Carolina State University,
Raleigh, North Carolina

Pergamon Press Inc.

New York · Toronto · Oxford · Sydney · Braunschweig

PERGAMON PRESS INC.
Maxwell House, Fairview Park, Elmsford, N.Y. 10523

PERGAMON OF CANADA LTD.
207 Queen's Quay West, Toronto 117, Ontario

PERGAMON PRESS LTD.
Headington Hill Hall, Oxford

PERGAMON PRESS (AUST.) PTY. LTD.
Rushcutters Bay, Sydney, N.S.W.

PERGAMON GmbH
Burgplatz 1, Braunschweig

Library of Congress Cataloging in Publication Data

Murray, Raymond LeRoy, 1920-
    Nuclear energy.

    (Pergamon unified engineering series, 19)
        1.   Nuclear engineering.   2.   Atomic energy.
I.   Title.
TK9146.M87   621.48   74-8685
ISBN 0-08-018164-3
ISBN 0-08-018163-5   (pbk.)

Printed in the United States of America

To Quin

Contents

The Author

Raymond L. Murray (Ph.D. University of Tennessee) is professor in the Department of Nuclear Engineering at North Carolina State University at Raleigh. His professional interests are in nuclear reactor design analysis, reactor and radiation safety, and nuclear engineering education. His textbook *Introduction to Nuclear Engineering* is widely known. He serves as a consultant to the nuclear industry. Dr. Murray's career in the nuclear field began in 1942 with the Manhattan Project at Berkeley and continued at Oak Ridge. In 1950 he helped found the first university nuclear engineering programs. He is Fellow of the American Physical Society and of the American Nuclear Society and is a member of several other scientific and engineering societies.

Preface

The future of mankind is inextricable from nuclear energy. As the world population increases and eventually stabilizes, the demands for energy to assure adequate living conditions will severely tax available resources, especially those of fossil fuels. New and different sources of energy and methods of conversion will have to be explored and brought into practical use. The wise use of nuclear energy, based on understanding of both hazards and benefits, will be required to meet this challenge to existence.

This book is intended to provide a factual description of basic nuclear phenomena, to describe devices and processes that involve nuclear reactions, and to call attention to the problems and opportunities that are inherent in a nuclear age. It is designed for use by anyone who wishes to know about the role of nuclear energy in our society or to learn nuclear concepts for use in professional work.

In spite of the technical complexity of nuclear systems, students who have taken a one-semester course based on the book have shown a surprising level of interest, appreciation, and understanding. This response resulted in part from the selectivity of subject matter and from efforts to connect basic ideas with the "real world," a goal that all modern education must seek if we hope to solve the problems facing civilization.

The sequence of presentation proceeds from fundamental facts and principles through a variety of nuclear devices to the relation between nuclear energy and peaceful applications. Emphasis is first placed on energy, atoms and nuclei, and nuclear reactions, with little background required. The book then describes the operating principles of radiation equipment, nuclear reactors, and other systems involving nuclear processes, giving quantitative information wherever possible. Finally, attention is directed to the subjects of radiation protection, beneficial usage of radiation, and the connection between energy resources and human progress.

The author is grateful to Dr. Ephraim Stam for his many suggestions on technical content, to Drs. Claude G. Poncelet and Albert J. Impink, Jr. for their careful review, to Christine Baermann for her recommendations on style and clarity, and to Carol Carroll for her assistance in preparation of the manuscript.

Raleigh, North Carolina   RAYMOND L. MURRAY

Part I   Basic Concepts

In the study of the practical applications of nuclear energy, we must take account of the properties of individual particles of matter—their "microscopic" features—as well as the character of matter in its ordinary form, a "macroscopic" (large-scale) view. Examples of the small-scale properties are masses of atoms or nuclear particles, their effective sizes for interaction with each other, or the number of particles in a certain volume. The combined behavior of large numbers of individual particles is expressed in terms of properties such as mass density, charge density, electrical conductivity, thermal conductivity, and elastic constants. We continually seek consistency between the microscopic and macroscopic views.

Since all processes involve interactions of particles, it is necessary that we develop a background of understanding of the basic physical facts and principles that govern such interactions. In Part I, we shall examine the concept of energy, describe the models of atomic and nuclear structure, discuss radioactivity and nuclear reactions in general, review the ways radiation reacts with matter, and concentrate on two important nuclear processes—fission and fusion.

1
Energy and States of Matter

Our material world is composed of many substances distinguished by their chemical, mechanical, and electrical properties. They are found in nature in various physical states—the familiar solid, liquid, and gas, along with the ionic "plasma." However, the apparent diversity of kinds and forms of material is reduced by the knowledge that there are only a little over 100 distinct chemical elements and that the chemical and physical features of substances depend merely on the strength of force bonds between atoms.

In turn, the distinctions between the elements of nature arise from the number and arrangement of basic particles—electrons, protons, and neutrons. At both the atomic and nuclear levels, the structure of elements is determined by internal forces and energy.

1.1 FORCES AND ENERGY

There is a limited number of basic forces—gravitational, electrostatic, electromagnetic, and nuclear. Associated with each of these is the ability to do work. Thus energy in different forms may be stored, released, transformed, transferred, and "used" in both natural processes and man-made devices. It is often convenient to view nature in terms of only two basic entities—particles and energy. Even this distinction can be removed, since we know that matter can be converted into energy and vice versa.

Let us review some principles of physics needed for the study of the release of nuclear energy and its conversion into thermal and electrical

3

form. We recall that if a constant force F is applied to an object to move it a distance s, the amount of work done is the product Fs. As a simple example, we pick up a book from the floor and place it on a table. Our muscles provide the means to lift against the force of gravity on the book. We have done work on the object, which now possesses stored energy (potential energy), because it could do work if allowed to fall back to the original level. Now a force F acting on a mass m provides an acceleration a, given by Newton's law $F = ma$. Starting from rest, the object gains a speed v, and at any instant has energy of motion (kinetic energy) in amount $E_k = \frac{1}{2}mv^2$. For objects falling under the force of gravity, we find that the potential energy is reduced as the kinetic energy increases, but the sum of the two types remains constant. This is an example of the principle of conservation of energy. Let us apply this principle to a practical situation and perform some illustrative calculations.

As we know, falling water provides one primary source for generating electrical energy. In a hydroelectric plant, river water is collected by a dam and allowed to fall through a considerable height. The potential energy of water is thus converted into kinetic energy. The water is directed to strike the blades of a turbine, which turns an electric generator. The force of gravity on a mass of water is $F = mg$, where g is the acceleration of gravity, 9.8 m/sec².† At the top of the dam of height h, the potential energy is $E_p = Fh$ or mgh. For instance, each kilogram at level 50 meters has energy $(1)(9.8)(50) = 490$ joules (J). Ignoring friction effects, this amount of energy in kinetic form would appear at the bottom. The speed of the water would be $v = \sqrt{2E_k/m} = 31.3$ m/sec.

Energy takes on various forms, classified according to the type of force that is acting. The water in the hydroelectric plant experiences the force of gravity, and thus gravitational energy is involved. It is transformed into mechanical energy of rotation in the turbine, which then is converted to electrical energy by the generator. At the terminals of the generator, there is an electrical potential difference, which provides the force to move charged particles (electrons) through the network of the electrical supply system. The electrical energy may then be converted into mechanical energy as in motors, or into light energy as in lightbulbs, or into thermal energy as in electrically heated homes, or into chemical energy as in a storage battery.

The automobile also provides familiar examples of energy transformations. The burning of gasoline releases the chemical energy of the fuel in

---

†Most of the time metric units based on the kilogram (kg), meter (m), and second (sec) will be employed in preference to the British units.

the form of heat, part of which is converted to energy of motion of mechanical parts, while the rest is transferred to the atmosphere and highway. Electricity is provided by the car's generator for control and lighting. In each of these examples, energy is changed from one form to another, but is not destroyed. The conversion of heat to other forms of energy is governed by two laws, the first and second laws of thermodynamics. The first states that energy is conserved; the second specifies inherent limits on the efficiency of energy conversion.

Energy can be classified according to the primary source. We have already noted two sources of energy: falling water and the burning of the chemical fuel gasoline, which is derived from petroleum, one of the main fossil fuels. To these we can add solar energy, the energy from winds, tides, or other sea motion, and heat from within the earth. Finally, we have energy from nuclear reactions, i.e., the burning of nuclear fuel.

1.2 THERMAL ENERGY

Of special importance to us is thermal energy, as the form most readily available from the sun, from burning of ordinary fuels, and from the fission process. First we recall that a simple definition of the temperature of a substance is the number read from a measuring device such as a thermometer in intimate contact with the material. If energy is supplied, the temperature rises, e.g., energy from the sun warms the air during the day. Each material responds to the supply of energy according to its internal molecular or atomic structure, characterized on a macroscopic scale by the specific heat c. If an amount of thermal energy added to one gram of the material is Q, the temperature rise, ΔT, is Q/c. To raise the temperature of 1 g of water by one degree Celsius (1°C), the amount of energy required is one calorie, which is also 4.18 J; thus for water $c = 4.18$ J/g-°C. The calorie (cal) is a unit devised before the equivalence of heat and mechanical energy was known.

From our modern knowledge of the atomic nature of matter, we readily appreciate the idea that energy supplied to a material increases the motion of the individual particles of the substance. Temperature can thus be related to the average kinetic energy of the atoms. For example, in a gas such as air, the average energy of translational motion of the molecules $\bar{E}$ is directly proportional to the temperature T, through the relation $\bar{E} = \frac{3}{2} kT$, where k is Boltzmann's constant, 1.38×10^{-23} J/°K. (Note that the Kelvin scale has the same spacing of degrees as does the Celsius scale, but its zero is at -273°C.)

To gain an appreciation of molecules in motion, let us find the typical speed of oxygen molecules at room temperature 20°C or 293°K. The molecular weight is 32, and since one unit of atomic weight corresponds to 1.66×10^{-27} kg, the mass of the oxygen (O_2) molecule is 5.3×10^{-26} kg. Now

$$\bar{E} = \tfrac{3}{2}(1.38 \times 10^{-23})(293) = 6.1 \times 10^{-21} \text{ J},$$

and thus the speed is

$$v = \sqrt{2\bar{E}/m} = \sqrt{2(6.1 \times 10^{-21})/(5.3 \times 10^{-26})} \cong 479 \text{ m/sec}.$$

Closely related to energy is the physical entity *power*, which is the rate at which work is done. To illustrate, suppose that the flow of water in the hydroelectric plant were 2×10^6 kg/sec. The corresponding energy per second is $(2 \times 10^6)(490) = 9.8 \times 10^8$ J/sec. For convenience, the unit joules per second is called the watt (W). Our plant thus involves 9.8×10^8 W. We can conveniently express this in kilowatts (1 kW = 10^3 W) or megawatts (1 MW = 10^6 W).

A large variety of units is encountered in our reading. For energy, we have joule, calorie, Btu (British thermal unit), kilowatt-hour, and foot-pound; for power, we have watt, horsepower, and foot-pounds per second. Some useful conversion factors for units are listed below:

1 calorie = 4.18 joules
1 kilowatt-hour = 3.6×10^6 joules = 3413 Btu
1 horsepower = 746 watts = 550 ft-lb/sec

In dealing with forces and energy at the level of molecules, atoms, and nuclei, it is conventional to use another energy unit, the *electron-volt* (eV). Its origin is electrical in character, being the amount of energy that would be imparted to an electron (charge 1.60×10^{-19} coulombs) if it were accelerated through a potential difference of 1 volt. Since the work done on 1 coulomb would be 1 J, we see that 1 eV = 1.60×10^{-19} J. The unit is of convenient size for describing atomic reactions. For instance, to remove the one electron from the hydrogen atom requires 13.5 eV of energy. However, when dealing with nuclear forces, which are very much larger than atomic forces, it is preferable to use the million-electron-volt unit (MeV). To separate the neutron from the proton in the nucleus of heavy hydrogen, for example, requires an energy of about 2.2 MeV, i.e., 2.2×10^6 eV.

1.3 RADIANT ENERGY

Another form of energy is electromagnetic or radiant energy. We recall that this energy may be, released by heating of solids, as in the wire of a lightbulb, or by electrical oscillations, as in radio or television transmitters, or by atomic interactions, as in the sun. The radiation can be viewed in either of two ways—as a combination of electric and magnetic waves traveling through space, or as a compact moving uncharged particle—the photon, which is a bundle of pure energy, effectively having mass only by virtue of its motion. The choice of model—wave or particle—is dependent on the process under study. Regardless of its origin, all radiation can be characterized by its energy or frequency, which is related to speed and wavelength. Letting c be the speed of light, λ its wavelength and ν its frequency, we have $c = \lambda\nu$.† For example, if c in a vacuum is 3×10^8 m/sec, yellow light of wavelength 5.89×10^{-7} m has a frequency of 5.1×10^{14} sec^{-1}. X-rays and gamma rays are electromagnetic radiation arising from the interactions of atomic and nuclear particles, respectively, and are different from light from other sources only in the range of energy represented.

In order to appreciate the relation of states of matter, atomic and nuclear interactions, and energy, let us visualize an experiment in which we supply energy to a sample of water from a source of energy that is as large and as sophisticated as we wish. Thus we increase the degree of internal motion and eventually dissociate the material into its most elementary components. Suppose, Fig. 1.1, that the water is initially as ice at nearly absolute zero temperature, where water (H_2O) molecules are essentially at rest. As we add thermal energy to increase the temperature to 0°C or 32°F, molecular movement increases to the point where the ice melts to become liquid water, which can flow rather freely. To cause a change from the solid state to the liquid state, a definite amount of energy (the heat of fusion) is required. In the case of water, this latent heat is 80 cal/g. In the temperature range in which water is liquid, thermal agitation of the molecules permits some evaporation from the surface. At the boiling point, 100°C or 212°F at atmospheric pressure, the liquid turns into the gaseous form as steam. Again, energy is required to cause the change of state, with a heat of vaporization of 539 cal/g. Further heating, using special high temperature equipment, causes dissociation

---

†We shall have need of both Roman and Greek characters, identifying the latter by name the first time they are used, thus λ (lambda) and ν (nu). The reader must be wary of symbols used for more than one quantity.

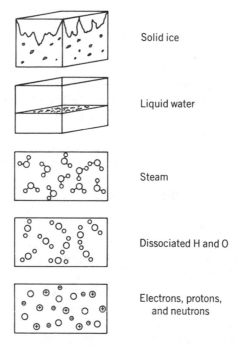

Solid ice

Liquid water

Steam

Dissociated H and O

Electrons, protons, and neutrons

Fig. 1.1. Effect of added energy.

of water into atoms of hydrogen (H) and oxygen (O). By electrical means, electrons can be removed from hydrogen and oxygen atoms, leaving a mixture of charged ions and electrons. Through nuclear bombardment, the oxygen nucleus can be broken into smaller nuclei, and in the limit of temperatures in the billions of degrees, the material can be decomposed into an assembly of electrons, protons, and neutrons.

1.4 THE EQUIVALENCE OF MATTER AND ENERGY

The connection between energy and matter is provided by Einstein's theory of special relativity. It predicts that the mass of any object increases with its speed. Letting the mass when the object is at rest be m_0, the "rest mass," and letting m be the mass when it is at speed v, and noting that the speed of light in a vacuum is $c = 3 \times 10^8$ m/sec, then

$$m = \frac{m_0}{\sqrt{1 - (v/c)^2}}.$$

For motion at low speed (e.g., 500 m/sec), the mass is almost identical to the rest mass, since v/c and its square are very small. Although the theory has the status of natural law, its rigor is not required except for particle motion at high speed, i.e., when v is at least a few percent of c. The relation shows that a material object can have a speed no higher than c.

The kinetic energy imparted to a particle by the application of force according to Einstein is

$$E_k = (m - m_0)c^2.$$

(For low speeds, $v \ll c$, this is approximately $\frac{1}{2}m_0v^2$, the classical relation.)

The implication of Einstein's formula is that any object has an energy $E_0 = m_0c^2$ when at rest (its "rest energy"), and a total energy $E = mc^2$, the difference being E_k the kinetic energy. Let us compute the rest energy for an electron of mass 9.1×10^{-31} kg.

$$E_0 = m_0c^2 = (9.1 \times 10^{-31})(3.0 \times 10^8)^2 = 8.2 \times 10^{-14} \text{ J}$$

or

$$E_0 = \frac{8.2 \times 10^{-14} \text{ J}}{1.60 \times 10^{-13} \text{ J/MeV}} = 0.51 \text{ MeV}.$$

For one unit of atomic mass, 1.66×10^{-27} kg, which is close to the mass of a hydrogen atom, the corresponding energy is 931 MeV.

Thus we see that matter and energy are equivalent, with the factor c^2 relating the amounts of each. This suggests that matter can be converted into energy and that energy can be converted into matter. Although Einstein's relation is completely general, it is especially important in calculating the release of energy by nuclear means. We find that *the energy yield from a pound of nuclear fuel is more than a million times that from chemical fuel.* To prove this startling statement, we first find the result of complete transformation of a kilogram of matter into energy, viz. $(1 \text{ kg})(3.0 \times 10^8 \text{ m/sec})^2 = 9 \times 10^{16} \text{ J}$. The nuclear fission process, as one method of converting mass into energy, is relatively inefficient, since the "burning" of 1 kg of uranium involves the conversion of only 0.87 g of matter into energy. This corresponds to about 7.8×10^{13} J/kg of the uranium consumed. The enormous magnitude of this energy release can be appreciated only by comparison with the energy of combustion of a familiar fuel such as gasoline, 5×10^7 J/kg. The ratio of these numbers, 1.5×10^6, reveals the tremendous difference between nuclear and chemical energies.

1.5 SUMMARY

Associated with each basic type of force is an energy, which may be transformed to another form for practical use. The addition of thermal energy to a substance causes an increase in temperature and the amount of particle motion. Electromagnetic radiation arising from electrical devices, atoms, or nuclei may be considered as composed of waves or of photons. Matter can be converted into energy and vice versa, according to Einstein's formula $E = mc^2$. The energy of nuclear fission is millions of times as large as that from chemical reactions.

1.6 PROBLEMS

1.1. Find the kinetic energy of a basketball player of mass 75 kg as he moves down the floor at a speed of 8 m/sec.

1.2. Recalling the conversion formulas for temperature,

$$T(°C) = \frac{5}{9}[T(°F) - 32]$$

and

$$T(°F) = \frac{9}{5}T(°C) + 32,$$

convert each of the following: 68°F, 500°F, −273°C, 1000°C.

1.3. If the specific heat of iron is 0.106 cal/g-°C, how many calories are required to bring 1 lb (454 g) of iron from 0°C to 100°C?

1.4. Find the speed corresponding to the average energy of nitrogen gas molecules (N_2, 28 units of atomic weight) at room temperature.

1.5. Find the power in kilowatts of an auto rated at 200 horsepower. In a drive for 4 hr at average speed 45 mph, how many kWhr of energy are required?

1.6. Find the frequency of a gamma ray photon of wavelength 1.5×10^{-12} m.

1.7. Verify that the mass of a typical slowly moving object is not much greater than its mass at rest (e.g., a car with $m_0 = 1000$ kg moving at 20 m/sec) by finding the number of *grams* of mass increase.

1.8. Noting that the electron-volt is 1.60×10^{-19} J, how many joules are released in the fission of one uranium nucleus, which yields 190 MeV?

1.9. Applying Einstein's formula for the equivalence of mass and energy, $E = mc^2$, where $c = 3 \times 10^8$ m/sec, the speed of light, how many kilograms of matter are converted into energy in Problem 1.8?

1.10. If the atom of uranium-235 has a mass of $(235)(1.66 \times 10^{-27})$ kg, what amount of equivalent energy does it have?

1.11. Using the results of Problems 1.8, 1.9, and 1.10, what fraction of the mass of a U-235 nucleus is converted into energy when fission takes place?

1.12. Show that to obtain a power of 1 W from fission of uranium, it is necessary to cause 3.3×10^{10} fission events per second.

2

Atoms and Nuclei

A complete understanding of the microscopic structure of matter and the exact nature of the forces acting is yet to be realized. However, excellent models have been developed to predict behavior to an adequate degree of accuracy for most practical purposes. These models are descriptive or mathematical, often based on analogy with large-scale processes, on experimental data, or on advanced theory.

2.1 ATOMIC THEORY

The most elementary concept is that matter is composed of individual particles—atoms—that retain their identity as elements in ordinary physical and chemical interactions. Thus a collection of helium atoms that forms a gas has a total weight that is the sum of the weights of the individual atoms. Also, when two elements combine to form a compound (e.g., if carbon atoms combine with oxygen atoms to form carbon monoxide molecules), the total weight of the new substance is the sum of the weights of the original elements.

There are more than 100 known elements. Most are found in nature; some are artificially produced. Each is given an atomic number in the periodic table of the elements—examples are hydrogen (H) 1, helium (He) 2, oxygen (O) 8, and uranium (U) 92. The symbol Z is given to the atomic number, which is also the number of electrons in the atom and determines its chemical properties.

Generally, the higher an element is in the periodic table the greater is its weight, either as an individual particle or as an assembly of particles on the standard scale. The atomic weights, labeled M, of the example elements above are approximately H 1.008, He 4.003, O 16.00, and U 238.0. These values represent the number of grams of the element in a sample that contains a specific number of particles—Avogadro's number, N_a, 6.02×10^{23}. We can easily find the number of atoms per cubic centimeter in a substance if its density ρ in grams per cubic centimeter is known. For example, if we had a container of helium gas with density $0.00018 \, g/cm^3$, each cubic centimeter would contain a fraction 0.00018/4.003 of Avogadro's number of helium atoms, i.e., 2.7×10^{19}. This procedure can be expressed as a convenient formula for finding N, the number per cubic centimeter for any material:

$$N = \frac{\rho}{M} N_a.$$

Thus in uranium with density $19 \, g/cm^3$, we find $N = (19/238)(6.02 \times 10^{23}) = 0.048 \times 10^{24} \, cm^{-3}$. The relation holds for compounds as well, if M is taken as the molecular weight. In water, H_2O, with $\rho = 1.0 \, g/cm^3$ and $M = 2(1.008) + 16.00 \cong 18.0$, we have $N = (1/18)(6.02 \times 10^{23}) = 0.033 \times 10^{24} \, cm^{-3}$. (The use of numbers times 10^{24} will turn out to be convenient later.)

2.2 GASES

Substances in the gaseous state are described approximately by the perfect gas law, relating pressure, volume, and absolute temperature,

$$pV = nkT,$$

where n is the number of particles and k is Boltzmann's constant. Increases in the temperature of the gas due to heating cause greater molecular motion, which results in an increase of particle bombardment of a container wall and thus of pressure on the wall. The particles of gas, each of mass m, have a variety of speeds v in accord with Maxwell's gas theory, as shown in Fig. 2.1. The most probable speed, at the peak of this maxwellian distribution, is dependent on temperature according to the relation

$$v_p = \sqrt{\frac{2kT}{m}}.$$

The kinetic theory of gases provides a basis for calculating properties such as the specific heat. Using the fact from Chapter 1 that the average

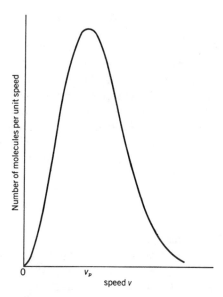

Fig. 2.1. Distribution of molecular speeds.

energy of gas molecules is proportional to the temperature, $\bar{E} = \frac{3}{2}kT$, we can deduce that the specific heat of a gas consisting only of atoms is $c = \frac{3}{2}k/m$, where m is the mass of one atom. We thus see an intimate relationship between mechanical and thermal properties of materials.

2.3 THE ATOM AND LIGHT

It is well known that the color of a heated solid or gas changes as the temperature is increased, tending to go from the red end of the visible region toward the blue end, i.e., from long wavelengths to short wavelengths. The measured distribution of light among the different wavelengths at a certain temperature can be explained by the assumption that light is in the form of photons. These are absorbed and emitted with definite amounts of energy E that are proportional to the frequency ν, according to

$$E = h\nu,$$

where h is Planck's constant 6.63×10^{-34} J-sec. For example, the energy corresponding to a frequency of 5×10^{14} is $(6.63 \times 10^{-34})(5 \times 10^{14}) = 3.3 \times 10^{-19}$ J, which is seen to be a very minute amount of energy.

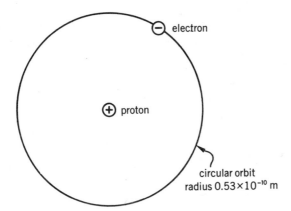

Fig. 2.2. Hydrogen atom.

The emission and absorption of light from incandescent hydrogen gas was first explained by Bohr, using a novel model of the hydrogen atom. He assumed that the atom consisted of a single electron moving at constant speed in a circular orbit about a nucleus—the proton—, as sketched in Fig. 2.2. Each particle has an electric charge of 1.6×10^{-19} coulombs, but the proton has a mass that is 1836 times that of the electron. The radius of the orbit is set by the equality of electrostatic force, attracting the two charges toward each other, to centripetal force, required to keep the electron on a circular path. If energy is supplied to the hydrogen atom from the outside, the electron is caused to jump to a larger orbit of definite radius. At some later time, the electron falls back spontaneously to the original orbit, and energy is released in the form of a photon of light. The energy of the photon $h\nu$ is equal to the difference between energies in the two orbits. The smallest orbit has a radius $R_1 = 0.53 \times 10^{-10}$ m, while the others have radii increasing as the square of integers (called quantum numbers). Thus if n is $1, 2, 3, \ldots$, the radius of the nth orbit is $R_n = n^2 R_1$. Figure 2.3 shows the allowed electron orbits in hydrogen. The energy of the atom system when the electron is in the first orbit is $E_1 = -13.5$ eV, where the negative sign means that energy must be supplied to remove the electron to a great distance and leave the hydrogen as a positive ion. The energy when the electron is in the nth orbit is $E_n = E_1/n^2$. The various discrete levels are sketched in Fig. 2.4.

The electronic structure of the other elements is described by the shell model, in which a limited number of electrons can occupy a given orbit or

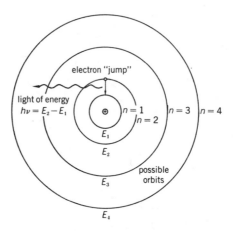

Fig. 2.3.   Electron orbits in hydrogen (Bohr theory).

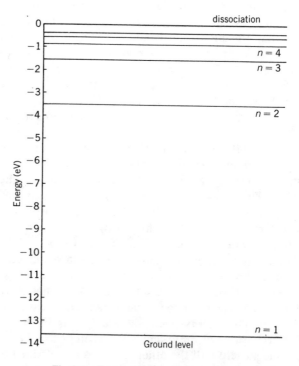

Fig. 2.4.   Energy levels in hydrogen atom.

shell. The atomic number Z is unique for each chemical element, and represents both the number of positive charges on the central massive nucleus of the atom and the number of electrons in orbits around the nucleus. The maximum allowed numbers of electrons in orbits as Z increases for the first few shells are 2, 8, and 18. The chemical behavior of elements is determined by the number of electrons in the outermost or valence shell. For example, oxygen with $Z = 8$ has two electrons in the inner shell, six in the outer. Thus oxygen has an affinity for elements with two electrons in the valence shell. The formation of molecules from atoms by electron sharing is illustrated by Fig. 2.5, which shows the water molecule.

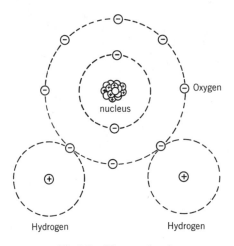

Fig. 2.5.   Water molecule.

2.4 NUCLEAR STRUCTURE

Most elements are composed of particles of different weight, called isotopes. For instance, hydrogen has three isotopes of weights in proportion 1, 2, and 3—ordinary hydrogen, heavy hydrogen (deuterium), and tritium. Each has atomic number $Z = 1$ and the same chemical properties, but they differ in the composition of the central nucleus, where most of the weight resides. The nucleus of ordinary hydrogen is the positively charged proton; the deuteron consists of a proton plus a neutron, a neutral particle of weight very close to that of the proton; the triton contains a proton plus two neutrons. To distinguish isotopes, we

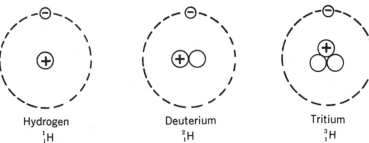

Fig. 2.6.   Isotopes of hydrogen.

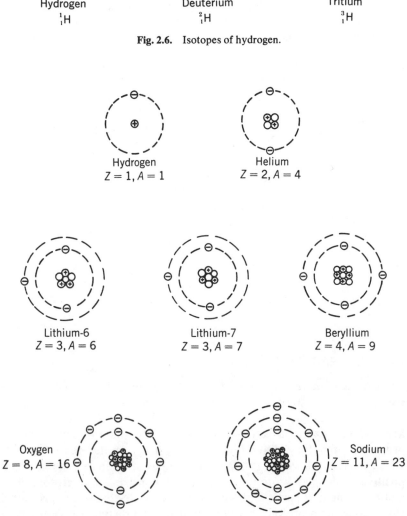

Fig. 2.7.   Atomic and nuclear structure.

identify the mass number A, as the total number of nucleons, the heavy particles in the nucleus. A complete shorthand description is given by the chemical symbol with superscript A value and subscript Z value, e.g., ${}^1_1H$, ${}^2_1H$, ${}^3_1H$. Figure 2.6 shows the nuclear and atomic structure of the three hydrogen isotopes. Each has one electron in the outer shell, in accord with the Bohr theory described earlier.

The structure of some of the lighter elements and isotopes is sketched in Fig. 2.7. In each case, the atom is neutral because the negative charge of the Z electrons in the outer shell balances the positive charge of the Z protons in the nucleus. The symbols for these are ${}^1_1H$, ${}^4_2He$, ${}^6_3Li$, ${}^7_3Li$, ${}^9_4Be$, ${}^{16}_8O$, and ${}^{23}_{11}Na$. In addition to the atomic number Z and the mass number A, we often need to write the neutron number N, which is, of course, $A - Z$. For the set of isotopes listed, N is 0, 2, 3, 4, 5, 8, and 12, respectively.

When we study nuclear reactions, it is convenient to let the neutron be represented by the symbol ${}^1_0n$, implying a mass comparable to that of hydrogen ${}^1_1H$, but with no electronic charge, $Z = 0$. Similarly, the electron is represented by ${}^0_{-1}e$, suggesting nearly zero mass in comparison with that of hydrogen, but with negative charge. An identification of isotopes frequently used in qualitative discussion consists of the element name and its A value, thus sodium-23 and uranium-235, or even more simply Na-23 and U-235.

2.5 SIZES AND MASSES OF NUCLEI

The dimensions of nuclei are found to be very much smaller than those of atoms. Whereas the hydrogen atom has a radius of about 5×10^{-9} cm, its nucleus has a radius of only about 10^{-13} cm. Since the proton weight is much larger than the electron weight, the nucleus is extremely dense. The nuclei of other isotopes may be viewed as closely packed particles of matter—neutrons and protons—forming a sphere whose volume, $\frac{4}{3}\pi R^3$, depends on A, the number of nucleons. A useful rule of thumb to calculate radii of nuclei is

$$R\,(\text{cm}) = 1.4 \times 10^{-13}\,A^{1/3}.$$

Since A ranges from 1 to about 250, we see that all nuclei are smaller than 10^{-12} cm.

The masses of atoms labeled M are compared on a scale in which an isotope of carbon ${}^{12}_6C$ has a mass of exactly 12. For ${}^1_1H$, the atomic mass is $M = 1.007825$, for ${}^2_1H$, $M = 2.014102$, and so on. The atomic mass of the proton is 1.007277, of the neutron 1.008665, the difference being only about 0.1%. The mass of the electron on this scale is 0.000549.

The atomic mass unit (amu), as $\frac{1}{12}$ the mass of $^{12}_{6}C$, corresponds to an actual mass of 1.66×10^{-24} g. To verify this, merely divide 1 g by Avogadro's number 6.02×10^{23}. It is easy to show that 1 amu is also equivalent to 931 MeV. We can calculate the actual masses of atoms and nuclei by multiplying the mass in atomic mass units by the mass of 1 amu. Thus the mass of the neutron is $(1.008665)(1.66 \times 10^{-24}) = 1.67 \times 10^{-24}$ g.

2.6 BINDING ENERGY

The force of electrostatic repulsion between like charges, which varies inversely as the square of their separation, would be expected to be so large that nuclei could not be formed. The fact that they do exist is evidence that there is an even larger force of attraction. This nuclear force acts only when the nucleons are very close to each other and binds them into a compact stable structure. Associated with the net force is a potential energy of binding. To disrupt a nucleus and separate it into its component nucleons, energy must be supplied from the outside. Recalling Einstein's relation between mass and energy, this is the same as saying that mass must be supplied to the nucleus. A given nucleus is lighter than the sum of its separate nucleons, the difference being the binding mass-energy. Let the mass of an atom including nucleus and external electrons be M, and let m_n and m_H be the masses of the neutron and the proton plus matching electron. Then the binding energy is

$$B = \text{total mass of separate particles} - \text{mass of the atom}$$

or

$$B = Nm_n + Zm_H - M.$$

(Neglected in this relation is a small energy of atomic or chemical binding.) Let us calculate B for tritium, the heaviest hydrogen atom. Figure 2.8 shows the dissociation that would taken place if sufficient energy were provided. Now $Z = 1$, $N = 2$, $m_n = 1.00866$, $m_H = 1.00782$, and $M = 3.01702$. Then

$$B = 2(1.00866) + 1(1.00782) - 3.01702$$

or

$$B = 0.00812 \text{ amu}.$$

Converting by use of the relation 1 amu = 931 MeV, the binding energy is $B = 7.56$ MeV. Calculations such as these are required for several purposes—to compare the stability of one nucleus with that of another, to find the energy release in a nuclear reaction, and to predict the possibility of fission of a nucleus.

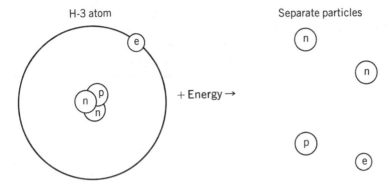

Fig. 2.8. Dissociation of tritium.

We can speak of the binding energy associated with one particle such as a neutron. Suppose that M_1 is the mass of an atom and M_2 is its mass after absorbing a neutron. The binding energy of the neutron of mass m_n is then

$$B_n = M_1 + m_n - M_2.$$

2.7 SUMMARY

All material is composed of elements whose chemical interaction depends on the number of electrons (Z). Light is absorbed and emitted in the form of photons when atomic electrons jump between orbits. Isotopes of elements differ according to the number of nucleons in the nucleus (A). Nuclei are much smaller than atoms and contain most of the mass of the atom. The nucleons are bound together by a net force in which the nuclear attraction forces exceeds the electrostatic repulsion forces. Energy must be supplied to dissociate a nucleus into its components.

2.8 PROBLEMS

2.1. Find the number of carbon ($^{12}_6C$) atoms in $1\,cm^3$ of graphite, density $1.65\,g/cm^3$.

2.2. Calculate the most probable speed of a "neutron gas" at temperature 20°C (293°K), noting that the mass of a neutron is $1.67 \times 10^{-27}\,kg$.

2.3. What frequency of light is emitted when an electron jumps into the smallest orbit of hydrogen, coming from a very large radius (assume infinity)?

2.4. Calculate the energy in electron-volts of the electron orbit in hydrogen for which $n = 3$, and find the radius in centimeters. How much energy would be needed to cause an electron to go from the innermost orbit to this one? If the electron jumped back, what frequency of light would be observed?

2.5. Sketch the atomic and nuclear structure of carbon-14, noting Z and A values and the numbers of electrons, protons, and neutrons.

2.6. What is the radius of the nucleus of uranium-238 viewed as a sphere? What is the area of the nucleus, seen from a distance as a circle?

2.7. Find the binding energy in MeV of ordinary helium $^4_2$He, for which $M = 4.002604$.

3

Radioactivity

Many naturally occurring and man-made isotopes have the property of radioactivity, which is the spontaneous disintegration (decay) of the nucleus with the emission of a particle. The process takes place in minerals of the ground, in fibers of plants, in tissues of animals, and in the air and water, all of which contain traces of radioactive elements.

3.1 NATURAL RADIOACTIVITY

Many heavy elements are radioactive. An example is the decay of the main isotope of uranium, in the reaction

$$^{238}_{92}U \rightarrow ^{234}_{90}Th + ^{4}_{2}He.$$

The particle released is the α (alpha) particle, which is merely the helium nucleus. The new isotope of thorium is also radioactive, according to

$$^{234}_{90}Th \rightarrow ^{234}_{91}Pa + ^{0}_{-1}e + \nu.$$

The three products are respectively the element protactinium, a β (beta) particle, which is merely an electron, and a neutrino symbolized by ν (nu). The latter is a neutral particle of zero rest mass that shares the reaction's energy release with the β particle. On the average, the neutrino carries $\frac{2}{3}$ of the energy, the electron, $\frac{1}{3}$. We note that the A value decreases by 4 and the Z value by 2 on emission of an α particle, while the A remains unchanged but Z increases by 1 on emission of a β particle. These two events are the start of a long sequence or "chain" of disintegrations that involve isotopes of the elements radium, polonium, and bismuth, eventually yielding the stable lead isotope $^{206}_{82}Pb$. Other chains found in nature start with $^{235}_{92}U$ and $^{232}_{90}Th$. Hundreds of "artificial" radioisotopes have been produced by bombardment of nuclei by charged particles or neutrons, and by separation of the products of the fission process.

3.2 THE DECAY LAW

The rate at which a radioactive substance disintegrates (and thus the rate of release of particles) depends on the isotopic species, but there is a definite "decay law" that governs the process. In a given time period, say one second, each nucleus of a given isotopic species has the same chance of decay. If we were able to watch one nucleus, it might decay in the next instant, or a few days later, or even hundreds of years later. Such statistical behavior is described by means of the *half-life* t_H, which is the time required for half of the nuclei to decay. We should like to know how many nuclei of a radioactive species remain at any time. If we start at time

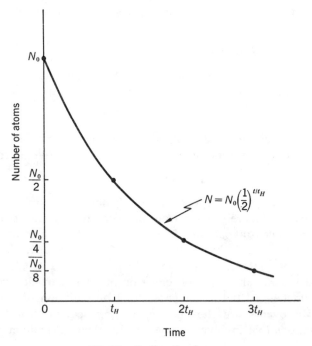

$$N = N_0\left(\frac{1}{2}\right)^{t/t_H}$$

Fig. 3.1. Radioactive decay.

zero with N_0 nuclei, after a length of time t_H, there will be $N_0/2$; by the time $2t_H$ has elapsed, there will be $N_0/4$; etc. A graph of the number of nuclei as a function of time is shown in Fig. 3.1. For any time t on the curve, the ratio of the number of nuclei present to the initial number is given by

$$\frac{N}{N_0} = \left(\frac{1}{2}\right)^{t/t_H}.$$

Half-lives range from very small fractions of a second to billions of years, with each isotope having a definite half-life. Table 3.1 gives several examples of radioactive materials with their emissions, product isotopes, and half-lives. The β particle energies are maximum values; on the average the emitted betas have only one-third as much energy. Included in the table are both natural and man-made radioactive isotopes (also called radioisotopes). We note the special case of neutron decay according to

$$\text{neutron} \rightarrow \text{proton} + \text{electron}.$$

A free neutron has a half-life of 11.7 min. The conversion of a neutron into a proton can be regarded as the origin of beta emission in radioactive nuclei. Most of the radioisotopes in nature are heavy elements. One exception is potassium-40, half-life 1.26×10^9 yr, with abundance 0.118%

Table 3.1. Selected Radioactive Isotopes.†

Isotope	Half-life	Principal Radiations (type, energy in MeV)
Neutron	11.7 min	β, 0.78
Hydrogen-3 (tritium)	12.262 yr	β, 0.0186
Carbon-14	5730 yr	β, 0.156
Sodium-24	14.96 hr	β, 1.389; γ, 1.369, 2.754
Phosphorus-32	14.28 day	β, 1.710
Potassium-40	1.26×10^9 yr	β, 1.314
Argon-41	1.83 hr	β, 1.198; γ, 1.293
Cobalt-60	5.263 yr	β, 0.314; γ, 1.173, 1.332
Krypton-85	10.76 yr	β, 0.67; γ, 0.514
Strontium-90	27.7 yr	β, 0.546
Iodine-131	8.05 day	β, 0.606
Radium-226	1602 yr	α, 4.78
Uranium-235	7.1×10^8 yr	α, 4.40
Uranium-238	4.51×10^9 yr	α, 4.20
Plutonium-239	24,390 yr	α, 5.16

†Reference: *Radiological Health Handbook*, Rockville, Md.: Public Health Service (1970).

in natural potassium. Others are carbon-14 and hydrogen-3 (tritium), which are produced continuously in small amounts by natural nuclear reactions. All three radioisotopes are found in plants and animals.

In addition to the radioisotopes that decay by beta or alpha emission, there is a large group of artificial isotopes that decay by the emission of a positron, which has the same mass as the electron and an equal but positive charge. An example is sodium-22, which decays with 2.62 yr half-life into a neon isotope as

$$^{22}_{11}\text{Na} \rightarrow {}^{22}_{10}\text{Ne} + {}^{0}_{+1}\text{e}.$$

Whereas the electron (also called negatron) is a normal part of any atom, the positron is not, and eventually is annihilated by combination with an electron to produce photons, as will be discussed in Chapter 6.

A nucleus can get rid of excess internal energy by the emission of a gamma ray, but in an alternate process called internal conversion the energy is imparted directly to one of the atomic electrons, ejecting it from the atom. In an inverse process called K-capture, the nucleus spontaneously absorbs one of its own orbital electrons. Each of these processes is followed by the production of X-rays as the inner shell vacancy is filled.

The formula for N/N_0 is not very convenient for calculations except when t is some integer multiple of t_H. Defining the decay constant λ (lambda), as the chance of decay of a given nucleus each second, an equivalent *exponential relation*† for decay is

$$\frac{N}{N_0} = e^{-\lambda t}.$$

We find that $\lambda = 0.693/t_H$. To illustrate, let us calculate the ratio N/N_0 at the end of two years for cobalt-60, half-life 5.26 yr. This artificially produced radioisotope has many medical and industrial applications. The reaction is

$$^{60}_{27}\text{Co} \rightarrow {}^{60}_{28}\text{Ni} + {}^{0}_{-1}\text{e} + 2\gamma,$$

---

†If λ is the chance one nucleus will decay in a second, then the chance in a time interval dt is λdt. For N nuclei, the change in number of nuclei is

$$dN = -\lambda N dt.$$

Integrating, and letting the number of nuclei at time zero be N_0 yields the formula quoted. Note that if

$$e^{-\lambda t} - \left(\frac{1}{2}\right)^{t/t_H},$$

then

$$\lambda t = \frac{t}{t_H} \log_e 2 \text{ or } \lambda = (\log_e 2)/t_H.$$

where the gamma ray energies are 1.17 and 1.33 MeV and the maximum beta energy is 0.314 MeV. Using the conversion $1 \, yr = 3.16 \times 10^7 \, sec$, $t_H = 1.66 \times 10^8 \, sec$. Then $\lambda = 0.693/(1.66 \times 10^8) = 4.17 \times 10^{-9} \, sec^{-1}$, and since t is $6.32 \times 10^7 \, sec$, λt is 0.264 and $N/N_0 = e^{-0.264} = 0.77$.

The rate of release of radiation by a radioisotope is dependent on the *activity*, A, which is the number of disintegrations per second. Since the decay constant λ is the chance of decay each second, then with N nuclei present, the activity is

$$A = \lambda N.$$

Let us find the activity for a sample of Co-60 consisting of 1 microgram, 10^{16} atoms. Now $A = (10^{16})(4.17 \times 10^{-9}) = 4.17 \times 10^7$ disintegrations per second (d/sec).

A useful unit of activity is the curie (Ci), named for the French scientists who worked with radium. The curie is 3.7×10^{10} d/sec, which is an early measured value of the activity per gram of radium. Our cobalt sample has a "strength" of $(4.17 \times 10^7)/(3.7 \times 10^{10}) = 0.0011$ Ci or 1.1 mCi.

The half-life tells us how long it takes for half of the nuclei to decay, while a related quantity, the mean life, τ, (tau) is the average time elapsed for decay of an individual nucleus. It turns out that τ is $1/\lambda$ and thus equal to $t_H/0.693$. For Co-60, τ is 7.6 yr.

3.3 MEASUREMENT OF HALF-LIFE

Finding the half-life of an isotope provides part of its identification, needed for beneficial use or for protection against radiation hazard. Let us look at a method for measuring the half-life of a radioactive substance. As in Fig. 3.2, a detector that counts the number of particles striking it is placed near the source of radiation. From the number of counts observed in a known short time interval, the counting rate is computed. It is proportional to the rates of emission of particles or rays from the sample and thus to the activity A of the source. The process is repeated after an elapsed time for decay. The resulting values of activity are plotted on semilog graph paper as in Fig. 3.3, and a straight line drawn through the observed points. From any pairs of points on the line λ and $t_H = 0.693/\lambda$ can be calculated (see Problem 3.7). The technique may be applied to mixtures of two radioisotopes. After a long time has elapsed, only the isotope of longest half-life will contribute counts. By extending its graph linearly back in time, one can find the counts to be subtracted from the total to yield the counts from the isotope of shorter half-life.

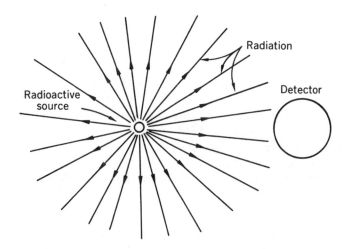

Fig. 3.2. Measurement of radiation from radioactive source.

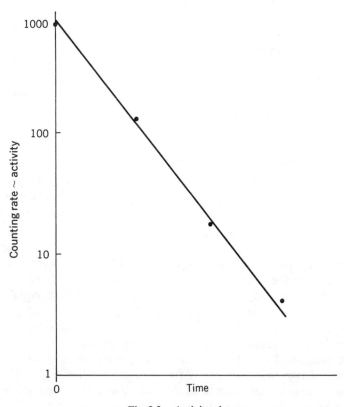

Fig. 3.3. Activity plot.

Activity plots cannot be used for a substance with very long half-life, e.g., strontium-90, 27.7 yr. The change in activity is almost zero over the span of time one is willing to devote to a measurement. However, if one knows the number of atoms present in the sample and measures the activity, the decay constant can be calculated from $\lambda = A/N$, from which t_H can be found.

The measurement of the activity of a radioactive substance is complicated by the presence of background radiation, which is due to cosmic rays from outside the earth or from the decay of minerals in materials of construction or in the earth. It is always necessary to measure the background counts and subtract them from those observed in the experiment.

3.4 SUMMARY

Many elements found in nature or man-made are radioactive, emitting alpha particles, beta particles, and gamma rays. The process is governed by an exponential relation, such that half of a sample decays in a time called the half-life t_H. Values of t_H range from fractions of a second to billions of years among the hundreds of radioisotopes known. Measurement of the activity, as the disintegration rate of a sample, yields half-life values, of importance in radiation use and protection.

3.5 PROBLEMS

3.1. Calculate the activity A for 1 g of radium-226 using the modern value of the half-life, and compare it with the definition of a curie.

3.2. The radioisotope sodium-24 ($^{24}_{11}Na$), half-life 15 hr, is used to measure the flow rate of salt water. By irradiation of stable $^{23}_{11}Na$ with neutrons, suppose that we produce 5 micrograms of the isotope. How much do we have at the end of 24 hr?

3.3. What was the initial and final activity in disintegrations per second and in curies for the sample of Na-24 in Problem 3.2?

3.4. The isotope uranium-238 ($^{238}_{92}U$) decays successively to form $^{234}_{90}Th$, $^{234}_{91}Pa$, $^{234}_{92}U$, $^{230}_{90}Th$, finally becoming radium-226 ($^{226}_{88}Ra$). What particles are emitted in each of these five steps? Draw a graph of this chain, using A and Z values on the horizontal and vertical axes, respectively.

3.5. A capsule of cesium-137, half-life 30 yr, is used to check the accuracy of detectors of radioactivity in air and water. Draw a graph on semilog paper of the activity over a 10-yr period of time, assuming the initial strength is 1 mCi. Explain the results.

3.6. A typical person has 140 g of potassium in his body. Of this a fraction

1.18×10^{-3} is K-40, half-life 1.26×10^{9} yr. How many disintegrations occur each second in the body?

3.7. (a) Noting that the activity of a radioactive substance is $A = \lambda N_0 e^{-\lambda t}$, verify that the graph of counting rate versus time on semilog paper is a straight line and show that

$$\lambda = \frac{\log_e (C_1/C_2)}{t_2 - t_1}$$

where points 1 and 2 are any pair on the curve.

(b) Using the following data, deduce the half-life of an "unknown," and suggest what isotope it is.

Time (sec)	Counting Rate (per sec)
0	200
1000	180
2000	162
3000	146
4000	131

3.8. By chemical means, we deposit 10^{-8} moles of a radioisotope on a surface and measure the activity to be 82,000 d/sec. What is the half-life of the substance and what element is it (see Table 3.1)?

4

Nuclear Reactions

Nuclear reactions are processes in which some change in the character of a nucleus takes place, either spontaneously as in radioactivity, or as the result of bombardment by a particle or ray.

4.1 ANALOGIES BETWEEN NUCLEAR AND CHEMICAL REACTIONS

There are many similarities between the two types of reactions, even though they occur in quite different energy regions and with different mechanisms. Individual particles are involved in each case—nucleons in nuclei and atoms in molecules, respectively. Conservation laws apply to each—that of charge, number of particles involved, and mass-energy. Reactions are similar in appearance. To illustrate, let us first write the simple equation describing the formation of water

$$2 H + O = H_2O.$$

The numbers of atoms of hydrogen and oxygen are the same on both sides; there is a valence balance, with $+1$ for hydrogen and -2 for oxygen; energy is released in the process, in this case 2.4 eV per molecule of water. Compare this with the reaction of a neutron with a proton, as the nucleus of hydrogen, to form a deuteron, as the nucleus of deuterium.

$$_0^1n + _1^1H \rightarrow _1^2H + \gamma.$$

The total number of nucleons is the same on both sides; the total nuclear charge is balanced; energy is released here in the form of a gamma ray, of about 2.2 MeV. We note that the energy release in this nuclear reaction is about a million times as great as for the chemical or atomic reaction.

A very large number of possible chemical reactions can take place because there are more than 100 known elements. Similarly, there are many nuclear reactions, involving about 2000 known isotopes, and several particles that can serve to induce reactions or be products of reaction—photons, electrons, protons, neutrons, alpha particles, deuterons, and heavier charged particles.

4.2 NUCLEAR TRANSMUTATION

The first example of artificial conversion of one element into another, a process of transmutation, was discovered by Rutherford in 1919. The reaction involved the bombardment of an isotope of nitrogen by alpha particles (nuclei of helium) from a radioactive source. The reaction products were an isotope of oxygen and a proton. The equation for the process is

$$_2^4\text{He} + {}_7^{14}\text{N} \rightarrow {}_8^{17}\text{O} + {}_1^1\text{H}.$$

We note that on both sides of the equation the sum of the A values is the same (18) and the sum of the Z values is the same (9). Figure 4.1 shows

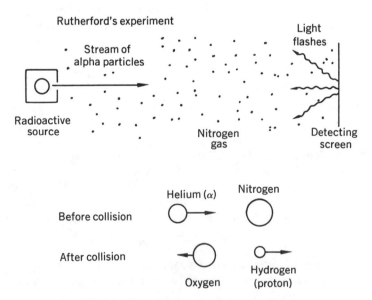

Fig. 4.1. Transmutation by nuclear reaction.

the reaction schematically. It is difficult for the alpha particle to enter the nitrogen nucleus because of the electrostatic repulsion of the two positively charged interacting nuclei. Thus the alpha particle must have several million-electron-volts of energy. Nuclear transmutations can also be induced by charged particles accelerated electrically to high speed. The first one discovered was

$$_1^1H + {}_3^7Li \rightarrow 2\,{}_2^4He.$$

Another reaction yielding a radioactive isotope of nitrogen is

$$_1^1H + {}_6^{12}C \rightarrow {}_7^{13}N + \gamma.$$

Nitrogen-13 emits a positron, the positive counterpart of the electron, with a half-life of 10 min.

4.3 NEUTRON REACTIONS

In contrast with charged particles, the neutron as a neutral particle need not have a high energy to penetrate the nucleus. Neutrons are thus especially effective as projectiles for inducing nuclear reactions. One of practically zero energy can be captured by hydrogen in the example described previously. That reaction yielded the stable isotope deuterium, normally present in nature. A radioactive isotope is produced by the reaction

$$_0^1n + {}_{27}^{59}Co \rightarrow {}_{27}^{60}Co + \gamma.$$

The cobalt-59 isotope has been changed into one of higher atomic mass, cobalt-60, which subsequently decays as discussed in Chapter 3. Note that energy is released instantly in the form of a capture gamma ray in the reaction that produces the radioisotope, and that energy is released much later in the form of decay gamma rays, beta particles, and neutrinos. Another example is neutron capture in cadmium, often used in control rods of a nuclear reactor, according to

$$_0^1n + {}_{48}^{113}Cd \rightarrow {}_{48}^{114}Cd + \gamma.$$

A reaction that may some day be employed to produce tritium, one of the fuels for a controlled thermonuclear reactor, is

$$_0^1n + {}_3^6Li \rightarrow {}_1^3H + {}_2^4He.$$

The alchemist's dream was realized when the reaction to produce gold from mercury was first performed,

$$_0^1n + {}_{80}^{198}Hg \rightarrow {}_{79}^{198}Au + {}_1^1H.$$

4.4 ENERGY BALANCES

The conservation of mass-energy is a firm requirement for any valid nuclear reaction. Recall from Chapter 1 that the total energy is made up of the rest energy (corresponding to a particle's mass at rest), and the kinetic energy of motion. The total energy of the reactants must equal that of the products. Strictly speaking, we should account for all of the effects of special relativity, including the mass increase with particle speed. Fortunately, however, for reactions of particles that come together and leave each other at low speeds we can use the classical formula $E_k = \frac{1}{2} m_0 v^2$ for the kinetic energy and the inherent energy associated with the rest mass $E_0 = m_0 c^2$.

Let us calculate the energy release from the neutron–proton reaction, assuming that the neutron is so slow that its kinetic energy can be neglected. Write a balance statement

mass of neutron + mass of hydrogen atom =
mass of deuterium atom + kinetic energy of products.

The kinetic energy is that of the $^2_1$H atom plus that of the gamma ray, the latter having no rest mass. We insert the accurately known masses of the particles,
$$1.008665 + 1.007825 = 2.014102 + E_k.$$

Solving, E_k is 0.002378 amu, and since 1 amu = 931 MeV, the energy release is 2.22 MeV.

We can further illustrate the energy balance idea by finding the energy of the two alpha particles released in the reaction of a proton with lithium-7. Suppose that the proton has a kinetic energy of 2 MeV which corresponds to $2/931 = 0.00215$ amu and that the target nucleus is at rest. The energy balance statement is

kinetic energy of hydrogen + mass of hydrogen + mass of lithium =
mass of helium + kinetic energy of helium
$$0.00215 + 1.00782 + 7.01600 = 2(4.00260) + E_k.$$

Then $E_k = 0.02077$ amu = 19.3 MeV. This energy is shared by the two alpha particles.

4.5 MOMENTUM CONSERVATION

The calculations just completed tell us the total kinetic energy of the product particles but do not say how much each has or what their speeds are. To find the kinetic energies of each of the particles we must apply the principle of conservation of momentum. Recall that the linear momentum p of a material particle of mass m and speed v is $p = mv$. This relation is correct in both the classical and relativistic senses. The total momentum of the interacting particles before and after the reaction is the same.

For our problem of a very slow neutron striking a hydrogen atom at rest, we can assume the initial momentum is zero. If it is to be zero finally, the ${}_1^2H$ and γ-ray must fly apart with equal magnitudes of momentum $p_d = p_\gamma$. The momentum of a gamma ray having the speed of light c may be written $p_\gamma = mc$ if we regard the mass as an *effective* value, related to the gamma energy E_γ by Einstein's formula $E = mc^2$. Thus

$$p_\gamma = \frac{E_\gamma}{c}.$$

Most of the 2.22 MeV energy release of the neutron capture reaction goes to the gamma ray, as shown in Problem 4.3. Assuming that to be correct, we can estimate the effective mass of this gamma ray. It is close to 0.00238 amu, which is very small compared with 2.014 amu for the deuterium. Then from the momentum balance, we see that the speed of recoil of the deuterium is very much smaller than the speed of light.

The calculation of the energies of the two alpha particles is a little complicated even for the case in which they separate along the same line that the proton entered. If we let m be the alpha particle mass and v_1 and v_2 be their speeds, with p_H the proton momentum, we must solve the two equations

$$mv_1 - mv_2 = p_H,$$
$$\tfrac{1}{2}mv_1^2 + \tfrac{1}{2}mv_2^2 = E_k.$$

4.6 GENERAL CONCEPTS OF NUCLEAR REACTIONS

A systematic description of nuclear reactions has been developed to account for experimental observations. Algebraic symbols are used to represent the particles involved in a sequence of events, as follows. Suppose that one particle labeled a strikes a nucleus X to produce a "compound nucleus" C^*, where the asterisk implies that the nucleus contains extra internal energy of motion of the nucleons, called excitation energy. The compound nucleus then disintegrates to release the particle b and a residual nucleus Y. In terms of reaction equations, this may be written

$$a + X \rightarrow C^*,$$
$$C^* \rightarrow Y + b,$$

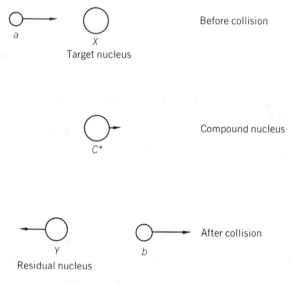

Fig. 4.2. General nuclear reaction.

with the net effect being

$$a + X = Y + b.$$

Figure 4.2 shows the event schematically. This reaction can be abbreviated as

$$X(a, b)Y,$$

which implies that particle a comes in and particle b goes out. The symbols a and b may stand for the neutron (n), alpha particle (α), deuteron (d), gamma ray (γ), proton (p), triton (t), the nucleus of tritium, and so on. In the previous sections, we have given examples of these reactions: (n, γ), (α, p), (p, α), (p, γ), (n, t), and (n, p).

For reactions at low energy, less than 10 MeV, the same compound nucleus could be formed by several different pairs of interacting nuclei, and could decay into several different pairs of final products. Consider Rutherford's transmutation reaction. The compound nucleus C^* in this case is the isotope $^{18}_{9}F^*$. It could have been formed in other ways, for instance by the combination of $^{1}_{1}H$ and $^{17}_{8}O$ or $^{1}_{0}n$ and $^{17}_{9}F$; it could disintegrate into $^{16}_{8}O$ and $^{2}_{1}H$ or $^{18}_{9}F$ and a gamma ray.

Depending on the binding energies of the nuclei, some of these reactions will release energy, others will require energy to be supplied. In the latter case, no nuclear reaction will take place if the energy of the incident particle is too low—the projectile will merely be scattered by the target. If the masses of all the particles involved are accurately known, the consequences of bombardment of a nucleus by any particle of any energy can be deduced, using the laws of conservation of energy and momentum.

4.7 SUMMARY

Chemical and nuclear reactions have several similarities in the form of equations and the requirements on conservation of particles and charge. The bombardment of nuclei by charged particles or neutrons produces new nuclei and particles. The final energies are found by taking account of mass difference, and the final speeds are obtained by applying the law of momentum conservation. In general, a compound nucleus having excitation energy is an intermediate step in a nuclear reaction.

4.8 PROBLEMS

4.1. The energy of formation of water from its constituent gases is quoted to be 54,500 cal/mole. Verify that this corresponds to 2.4 eV per molecule of H_2O.

4.2. Complete the following nuclear reaction equations:

$$\begin{align} {}^{1}_{0}n + {}^{14}_{7}N &\rightarrow \{\ \}(\quad) + {}^{1}_{1}H, \\ {}^{2}_{1}H + {}^{9}_{4}Be &\rightarrow \{\ \}(\quad) + {}^{1}_{0}n. \end{align}$$

4.3. Using the accurate atomic masses listed below, find the minimum amount of energy an alpha particle must have to cause the transmutation of nitrogen to oxygen. (${}^{14}_{7}N$ 14.003074, ${}^{4}_{2}He$ 4.002603, ${}^{17}_{8}O$ 16.999133, ${}^{1}_{1}H$ 1.007825.)

4.4. Find the energy release in the reaction ${}^{6}_{3}Li$ (n, α) ${}^{3}_{1}H$, noting the masses ${}^{0}_{1}n$ 1.00866, ${}^{3}_{1}H$ 3.01605, ${}^{4}_{2}He$ 4.00260, and ${}^{6}_{3}Li$ 6.01512.

4.5. A slow neutron of mass 1.00866 amu is captured by the nucleus of a hydrogen atom of mass 1.00782, and the final products are a deuterium atom of mass 2.01410 and a gamma ray. The energy released is 2.22 MeV. If the gamma ray is assumed to have almost all of this energy, what is its effective mass in kg? What is the speed of the ${}^{2}_{1}H$ particle in m/sec, using equality of momenta on separation? What is the recoil energy of ${}^{2}_{1}H$ in MeV? How does this compare with the total energy released? Was the assumption about the gamma ray reasonable?

4.6. Calculate the speeds and energies of the individual alpha particles in the reaction ${}^{1}_{1}H + {}^{7}_{3}Li \rightarrow 2\,{}^{4}_{2}He$, assuming that they separate along the line of proton motion. Note that the mass of the lithium-7 atom is 7.016004.

5

Reaction Rates

The chance of interaction of any two particles obviously depends on the nature of the force between them. The most familiar force is that of electrostatics, described by Coulomb's relation $F \sim q_1 q_2 / r^2$, where the q's are the charges carried by the objects and r is the distance of separation of their centers. Two charged particles, such as electrons or protons, will influence each other somewhat, no matter how far they are apart. On the other hand, two atoms, which are each neutral because of the equality of electronic and nuclear charge, will not interact appreciably until they get close to each other (around 10^{-8} cm). The special force between nucleons is similarly limited to small distances of separation (around 10^{-13} cm).

5.1 CROSS SECTIONS FOR PARTICLE INTERACTION

It is natural to think of a particle as having a definite size and, if viewed as a sphere, as having a radius. The question arises as to how such a small radius can be found. For objects of this size, the use of ordinary measuring devices such as a ruler is evidently impossible. Some form of microscope is required, using electromagnetic radiation—X-rays or gamma rays—or material particles—electrons, protons, or neutrons. However, the chance of interaction depends on the properties of both the particle used as probe and the particle under scrutiny. The radius deduced thus depends on the method of measurement. We must be satisfied with information on the apparent radius or the apparent cross-sectional area of target, each of which depends on the particular interaction. This leads to the concept of cross section, as a measure of the chance of collision between any two particles, whether they be atoms, nuclei, or photons.

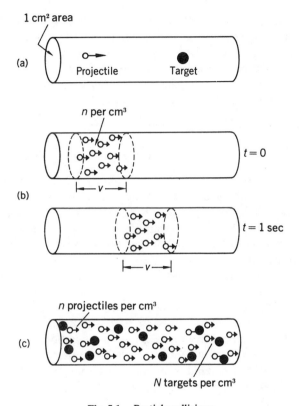

1 cm² area

(a)
Projectile   Target

n per cm³

(b)

$t = 0$

$\longmapsto v \longrightarrow$

$t = 1$ sec

$\longmapsto v \longrightarrow$

n projectiles per cm³

(c)

N targets per cm³

Fig. 5.1.   Particle collisions.

We can perform a set of imaginary experiments that will clarify the idea of cross section. Picture, as in Fig. 5.1a, a tube of end area 1 cm² containing only one target particle. A single projectile is injected parallel to the tube axis, but its exact location is not specified. It is clear that the chance of collision, labeled σ (sigma) and called the *microscopic cross section*, is the ratio of the target area to the area of the tube, which is 1.

Now let us inject a continuous stream of particles of speed v into the empty tube (see Fig. 5.1b). In a time of one second, each of the particles has moved along a distance v cm. All of them in a column of volume $(1 \text{ cm}^2)(v \text{ cm}) = v \text{ cm}^3$ will sweep past a point at which we watch each second. If there are n particles per cubic centimeter, then the number per unit time that cross any unit area perpendicular to the stream direction is nv, called the *current density*.

Finally, Fig. 5.1c, we fill each unit volume of the tube with N targets, each of area σ as seen by incoming projectiles (we presume that the targets do not "shadow" each other). If we focus attention on a unit volume, there is a total target area of $N\sigma$. Again, we inject the stream of projectiles. In a time of one second, the number of them that pass through the target volume is nv; and since the chance of collision of each with one target atom is σ, the number of collisions is $nvN\sigma$. We can thus define the reaction rate per unit volume,

$$R = nvN\sigma.$$

We let the current density nv be abbreviated by j and let the product $N\sigma$ be labeled Σ (capital sigma), the *macroscopic cross section*, referring to the large-scale properties of the medium. Then the reaction rate per cubic centimeter is simply $R = j\Sigma$. We can easily check that the units of j are $cm^{-2}\ sec^{-1}$ and those of Σ are cm^{-1}, so that the units of R are $cm^{-3}\ sec^{-1}$.

In a different experiment, we release particles in a medium and allow them to make many collisions with those in the material. In a short time, the directions of motion are random, as sketched in Fig. 5.2. We shall look only at particles of the same speed v, of which there are n per unit volume. The product nv in this situation is no longer called current density, but is given a different name, the flux, symbolized by ϕ (phi). If we place a unit area anywhere in the region, there will be flows of particles across it each second from both directions, but it is clear that the current densities will now be less than nv. It turns out that they are each

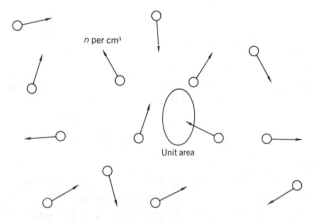

Fig. 5.2. Particles in random motion.

$nv/4$, and the total current density is $nv/2$. The rate of reaction of particles with those in the medium can be found by adding up the effects of individual projectiles. Each behaves the same way in interacting with the targets, regardless of direction of motion. The reaction rate is again $nvN\sigma$ or, for this random motion, $R = \phi\Sigma$.

When a particle such as a neutron collides with a target nucleus, there is a certain chance of each of several reactions. The simplest is elastic scattering, in which the neutron is visualized as bouncing off the nucleus and moving in a new direction with a change in energy. Such a collision, governed by classical physics, is predominant in light elements. In the inelastic scattering collision, an important process for fast neutrons in heavy elements, the neutron becomes a part of the nucleus, its energy provides excitation, and a neutron is released. The cross section σ_s is the chance of a collision that results in neutron scattering. The neutron may instead be absorbed by the nucleus, with cross section σ_a. Since σ_a and σ_s are chances of reaction, their sum is the chance for collision or total cross section $\sigma = \sigma_a + \sigma_s$.

Let us illustrate these ideas by some calculations. In a typical nuclear reactor used for training and research in universities, a large number of neutrons will be present with energies near 0.0253 eV. This energy corresponds to a most probable speed of 2200 m/sec for the neutrons viewed as a gas at room temperature, 293° absolute. Suppose that the flux of such neutrons is 2×10^{12} cm^{-2}-sec. The number density is then

$$n = \frac{\phi}{v} = \frac{2 \times 10^{12} \text{ cm}^{-2}\text{-sec}}{2.2 \times 10^5 \text{ cm/sec}} = 9 \times 10^6 \text{ cm}^{-3}.$$

Although this is a very large number by ordinary standards, it is exceedingly small compared with the number of water molecules per cubic centimeter (3.3×10^{22}) or even the number of air molecules per cubic centimeter (2.7×10^{19}). The "neutron gas" in a reactor is almost a perfect vacuum.

Now let the neutrons interact with uranium-235 as fuel in the reactor. The cross section for absorption σ_a is 681×10^{-24} cm^2. If the number density of fuel atoms is $N = 0.048 \times 10^{24}$ cm^{-3}, as in uranium metal, then the macroscopic cross section is

$$\Sigma_a = N\sigma_a = (0.048 \times 10^{24} \text{ cm}^{-3})(681 \times 10^{-24} \text{ cm}^2) = 32.7 \text{ cm}^{-1}.$$

The unit of area 10^{-24} cm^2 is conventionally called the *barn*.† If we express

†As the story goes, an early experimenter observed that the cross section for U-235 was "as big as a barn."

the number of targets per cubic centimeter in units of 10^{24} and the microscopic cross section in barns, then $\Sigma_a = (0.048)(681) = 32.7$ as above. With a neutron flux $\phi = 2 \times 10^{13}$ cm^{-2}-sec, the reaction rate for absorption is

$$R = \phi \Sigma_a = (2 \times 10^{13} \text{ cm}^{-2}\text{-sec})(32.7 \text{ cm}^{-1}) = 6.5 \times 10^{14} \text{ cm}^{-3}\text{-sec}.$$

This is also the rate at which uranium-235 nuclei are consumed.

The average energy of neutrons in a nuclear reactor used for electrical power generation is about 0.1 eV, almost four times the value used in our example. The effects of the high temperature of the medium (about 600°F) and of neutron absorption give rise to this higher value.

5.2 PARTICLE ATTENUATION

Visualize an experiment in which a stream of particles of common speed and direction is allowed to strike the plane surface of a substance as in Fig. 5.3. Collisions with the target atoms in the material will continually remove projectiles from the stream, which will thus diminish in strength with distance, a process we label attenuation. If the current density incident on the substance at position $z = 0$ is labeled j_0, the

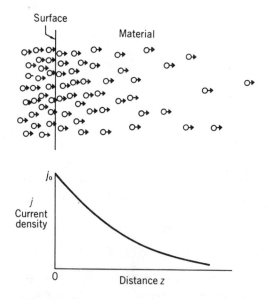

Fig. 5.3. Neutron penetration and attenuation.

current of those not having made any collision on penetrating to a depth z
is given by[†]

$$j = j_0 e^{-\Sigma z},$$

where Σ is the macroscopic cross section. The similarity in form to the
exponential for radioactive decay is noted, and one can deduce by
analogy that the half-thickness, the distance required to reduce j to half
its initial value, is $z_H = 0.693/\Sigma$. Another more frequently used quantity is
the mean free path λ, as the average distance a particle goes before
making a collision. By analogy with the mean life for radioactivity, we can
write[‡]

$$\lambda = \frac{1}{\Sigma}.$$

This relation is applicable as well to particles moving randomly in a
medium. Consider a particle that has just made a collision and moves off
in some direction. On the average, it will go a distance λ through the array
of targets before colliding again. For example, we can find the mean free
path for 1 eV neutrons in water, assuming that scattering by hydrogen
with cross section 20 barns is the dominant process. Now the number of
hydrogen atoms is $N_H = 0.0668 \times 10^{24} \, \text{cm}^{-3}$, σ_s is $20 \times 10^{-24} \, \text{cm}^2$, and
$\Sigma_s = 1.34 \, \text{cm}^{-1}$. Thus the mean free path for scattering λ_s is around
0.75 cm.

The cross sections for *atoms* interacting with their own kind at the
energies corresponding to room temperature conditions are of the order
of $10^{-15} \, \text{cm}^2$. If we equate this area to πr^2, the calculated radii are of the
order of 10^{-8} cm. This is in rough agreement with the theoretical radius of
electron motion in the hydrogen atom 0.53×10^{-8} cm. On the other hand,
the cross sections for *neutrons* interacting with nuclei by *scattering*
collisions, those in which the neutron is deflected in direction and loses
energy, are usually very much smaller than those for atoms. For the case
of 1 eV neutrons in hydrogen with a scattering cross section of 20 barns,
i.e., $20 \times 10^{-24} \, \text{cm}^2$, one deduces a radius of about 2.5×10^{-12} cm. These
results correspond to our earlier observation that the nucleus is thousands
of times smaller than the atom.

[†]The derivation proceeds as follows. In a slab of material of unit area and infinitesimal
thickness dz, the target area will be $N\sigma \, dz$. If the current at z is j, the number of collisions
per second in the slab is $jN\sigma \, dz$, and thus the change in j on crossing the layer is $dj = -j\Sigma \, dz$
where the reduction is indicated by the negative sign. By analogy with the solution of the
radioactive decay law, we can write the formula cited.

[‡]This relation can be derived directly by use of the definition of an average as the sum of
the distances the particles travel divided by the total number of particles. Using integrals,
this is $\bar{z} = \int z \, dj / \int dj$.

5.3 NEUTRON CROSS SECTIONS

The cross section for neutron *absorption* in materials depends greatly on the isotope bombarded and on the neutron energy. For consistent comparison and use, the cross section is often cited at 0.0253 eV, corresponding to neutron speed 2200 m/sec. Values for absorption cross sections for a number of isotopes at that energy are listed in order of increasing size in Table 5.1. The dependence of absorption cross section on energy is of two types, one called $1/v$, in which σ_a varies inversely with neutron speed, the other called resonance, where there is a very strong absorption at certain neutron energies. Many materials exhibit both variations. Figures 5.4 and 5.5 show the cross sections for boron and natural uranium. The use of the logarithmic plot enables one to display the large range of cross section over the large range of energy of interest.

Table 5.1.   Thermal Neutron Absorption Cross Sections.†

Isotope	σ_a (barns)
$^4_2$He	$\simeq 0$
$^{16}_8$O	0.000178
$^{12}_6$C	0.0034
$^2_1$H	0.0053
$_{40}$Zr	0.185
$^1_1$H	0.332
$^{238}_{92}$U	2.70
$^{235}_{92}$U	681
$^{239}_{94}$Pu	1011
$^{10}_5$B	3837
$^{135}_{54}$Xe	2,650,000

†Main source: BNL-325, *Neutron Cross Sections*, 3rd ed., Vol. 1, 1973.

Neutron scattering cross sections are more nearly the same for all elements and have less variation with neutron energy. Figure 5.6 shows the trend of σ_s for hydrogen as in water. Over a large range of neutron energy the scattering cross section is nearly constant, dropping off in the million-electron-volt region. This high energy range is of special interest since neutrons produced by the fission process have such energy values.

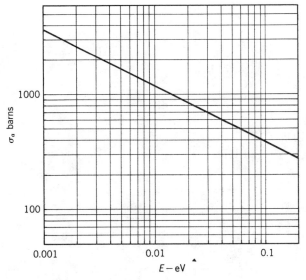

Fig. 5.4.   Absorption cross section for boron.

5.4   NEUTRON SLOWING

When fast neutrons, those of energy of the order of 2 MeV, are introduced into a medium, a sequence of collisions with nuclei takes place. The neutrons are deflected in direction on each collision, they lose energy, and they tend to migrate away from their origin. Each neutron has a unique history, and it is impractical to keep track of all of them. Instead, we seek to deduce average behavior. First, we note that the elastic scattering of a neutron with a nucleus of mass number A causes a reduction in neutron energy from E_0 to E and a change of direction through an angle θ (theta), as sketched in Fig. 5.7. The length of arrows indicates the speeds of the particles. This example shown is but one of a great variety of possible results of scattering collisions. For each final energy there is a unique angle of scattering, and vice versa, but the occurrence of a particular E and θ pair depends on chance. The neutron may bounce directly backward, $\theta = 180°$, dropping down to a minimum energy αE_0, where $\alpha = (A - 1)^2/(A + 1)^2$, or it may be undeflected, $\theta = 0°$, and retain its initial energy E_0, or it may be scattered through any other angle, with corresponding energy loss.

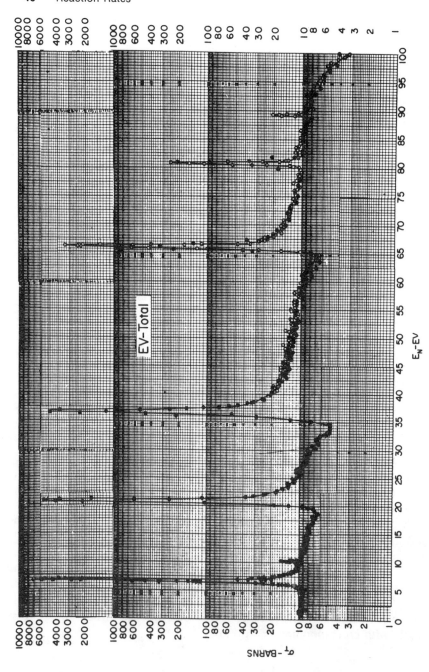

Fig. 5.5.   Cross section for natural uranium.

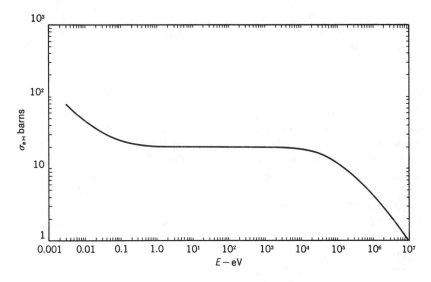

Fig. 5.6.   Scattering cross section for hydrogen.

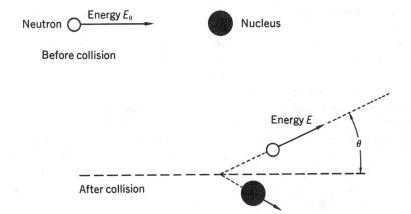

Fig. 5.7.   Neutron scattering and energy loss.

The average elastic scattering collision is described by two quantities that depend only on the nucleus, not on the neutron energy. The first is $\overline{\cos \theta}$, the average of the cosines of the angles of scattering, given by

$$\overline{\cos \theta} = \frac{2}{3A}.$$

For hydrogen, it is $\frac{2}{3}$, meaning that the neutron tends to be scattered in the forward direction; for a very heavy nucleus such as uranium, it is near zero, meaning that the scattering is almost equally likely in each direction. Forward scattering results in an enhanced migration of neutrons from their point of appearance in a medium. Their free paths are effectively longer, and it is conventional to use the "transport mean free path" $\lambda_t = \lambda_s/(1 - \overline{\cos \theta})$ instead of λ_s to account for the effect. We note that λ_t is always the larger. Consider slow neutrons in carbon, for which $\sigma_s = 4.8$ barns and $N = 0.083$, so that $\Sigma_s = 0.4 \text{ cm}^{-1}$ and $\lambda_s = 2.5$ cm. Now $\overline{\cos \theta} = 2/(3)(12) = 0.056$, $1 - \overline{\cos \theta} = 0.944$, and $\lambda_t = 2.5/0.944 = 2.7$ cm.

The second quantity that describes the average collision is ξ (xi), the average change in the natural logarithm of the energy, given by

$$\xi = 1 + \frac{\alpha \ln \alpha}{1 - \alpha}.$$

For hydrogen, it is exactly 1, the largest possible value, meaning that hydrogen is a good "moderator" for neutrons, its nuclei permitting the greatest neutron energy loss; for a heavy element it is $\xi \simeq 2/(A + \frac{2}{3})$ which is much smaller than 1, e.g., for carbon, $A = 12$, it is 0.16.

To find how many collisions C are required to slow neutrons from one energy to another, we merely divide the total change in $\ln E$ by ξ, the average per collision. In going from the fission energy 2×10^6 eV to the thermal energy 0.025 eV, the total change is $\ln (2 \times 10^6) - \ln (0.025) = \ln (8 \times 10^7) = 18.2$. Then $C = 18.2/\xi$. For example in hydrogen, $\xi = 1$, C is 18, while in carbon $\xi = 0.16$, C is 114. Again, we see the virtue of hydrogen as a moderator. The fact that hydrogen has a scattering cross section of 20 barns over a wide range while carbon has a σ_s of only 4.8 barns implies that collisions are more frequent and the slowing takes place in a smaller region. The only disadvantage is that hydrogen has a larger thermal neutron absorption cross section, 0.332 barns versus 0.0034 for carbon.

The movement of individual neutrons through a moderator during slowing consists of free flights, interrupted frequently by collisions that cause energy loss. Picture, as in Fig. 5.8, a fast neutron starting at a point,

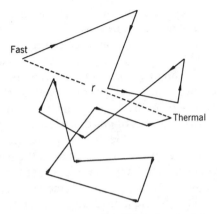

Fig. 5.8. Neutron migration during slowing.

and migrating outward. At some distance r away, it arrives at the thermal energy. Other neutrons become thermal at different distances, depending on their particular histories. If we were to measure all of their r values and form the average of r^2, the result would be $\overline{r^2} = 6\tau$, where τ is called the "age" of the neutron. Approximate values of the age for various reactor materials, as obtained from experiment are listed below

Moderator	τ, age to thermal (cm^2)
H_2O	26
D_2O	125
C	364

We thus note that water is a much better agent for neutron slowing than is graphite.

5.5 NEUTRON DIFFUSION

As neutrons slow into the energy region that is comparable to that of agitation of the moderator atoms, they may either lose or gain energy on collision. Members of a group of neutrons have various speeds at any instant and thus the group behaves as a gas, with temperature T that is close to that of the medium in which they are found. Thus if the moderator is at room temperature 20°C or 293°K, the most likely neutron speed is around 2200 m/sec, corresponding to a kinetic energy of 0.0253 eV. To a first approximation the neutrons have a maxwellian distribution comparable to that of a gas, as was shown in Fig. 2.1.

The process of diffusion of gas molecules is familiar to us. If a bottle of perfume is opened, the scent is quickly observed, as the molecules of the substance migrate away from the source. Since neutrons in large numbers behave as a gas, the descriptions of gas diffusion may be applied. The flow of neutrons through space at a location is proportional to the way the concentration of neutrons varies, in particular to the negative of the slope of the neutron number density. We can guess that the larger the neutron speed v and the larger the transport mean free path λ_t, the more neutron flow will take place. Theory and measurement show that if n varies in the z-direction, the net flow of neutrons across a unit area each second, the net current density, is

$$j = \frac{-\lambda_t v}{3} \frac{dn}{dz}.$$

This is called Fick's law of diffusion, derived long ago for the description of gases. It applies if absorption is small compared with scattering. In terms of the flux $\phi = nv$ and the *diffusion coefficient* $D = \lambda_t/3$, this may be written compactly $j = -D\phi'$, where ϕ' is the slope of the neutron flux.

5.6 SUMMARY

The cross section for interaction of neutrons with nuclei is a measure of chance of collision. The reaction rate per cubic centimeter is mutually dependent on current density (j) and macroscopic cross section (Σ). The stream formed by uncollided particles is reduced exponentially as it passes through a medium. Neutron absorption cross sections vary greatly with target isotope and with neutron energy, while scattering cross sections are almost constant. Neutrons are slowed readily by collisions with light nuclei and migrate from their point of origin. On reaching thermal energy they continue to disperse, and the net flow is proportional to the negative of the slope of the flux.

5.7 PROBLEMS

5.1. Calculate the macroscopic cross section for scattering of 1 eV neutrons in water, using N for water as 0.0334×10^{24} cm^{-3} and cross sections 20 barns for hydrogen and 3.8 barns for oxygen. Find the mean free path λ_s.

5.2. Find the speed v and the number density of neutrons of energy 1.5 MeV in a flux 7×10^{13} cm^{-2}-sec.

5.3. Compute the flux, macroscopic cross section and reaction rate for the following data: $n = 2 \times 10^5$ cm^{-3}, $v = 3 \times 10^8$ cm/sec, $N = 0.04 \times 10^{24}$ cm^{-3}, $\sigma = 0.5 \times 10^{-24}$ cm^2.

5.4. What are the values of the average logarithmic energy change ξ and the average cosine of the scattering angle $\overline{\cos \theta}$ for neutrons in beryllium, $A = 9$? How many collisions are needed to slow neutrons from 2 MeV to 0.025 eV in Be-9? What is the value of the diffusion coefficient D for 0.025 eV neutrons if Σ_s is 0.90 cm^{-1}?

5.5. (a) Verify that neutrons of speed 2200 m/sec have an energy of 0.0253 eV. (b) If the neutron absorption cross section of boron at 0.0253 eV is 759 barns, what would it be at 0.1 eV? Does this result agree with that shown in Fig. 5.4?

6

Radiation and Materials

The word "radiation" will be taken to embrace all particles, whether they are of material or electromagnetic origin. We include those produced by both atomic and nuclear processes and those resulting from electrical acceleration, noting that there is no essential difference between X-rays from atomic collisions and gamma rays from nuclear decay; protons can come from a particle accelerator, from cosmic rays, or from a nuclear reaction in a reactor. The word "materials" will refer to bulk matter, whether of mineral or biological origin, as well as the particles of which the matter is composed, including molecules, atoms, electrons, and nuclei.

When we put radiation and materials together, a great variety of possible situations must be considered. Bombarding particles may have low or high energy; they may be charged, uncharged, or photons; they may be heavy or light in the scale of masses. The targets may be similarly distinguished, but also exhibit degrees of binding that range from none ("free" particles), to weak (atoms in molecules and electrons in atoms), to strong (nucleons in nuclei). In most interactions, the higher the projectile energy in comparison with the energy of binding of the structure, the greater is the effect.

Out of the broad subject we shall select for review some of the reactions that are important in the nuclear energy field. Looking ahead, we shall need to understand the effects produced by the particles and rays from radioactivity and other nuclear reactions. Materials affected may be in or around a nuclear reactor, as part of its construction or inserted to be irradiated. Materials may be of biological form, including the human body, or they may be inert substances used for protective shielding against radiation. We shall not attempt to explain the processes rigorously, but be content with qualitative descriptions based on analogy with collisions viewed on an elementary physics level.

6.1 EXCITATION AND IONIZATION BY ELECTRONS

These processes occur in the familiar fluorescent lightbulb, or in a vacuum tube used in electrical devices, in an X-ray machine, or in matter exposed to beta particles. If an electron that enters a material has a very low energy, it will merely migrate without affecting the molecules significantly. If its energy is larger, it may impart energy to atomic electrons as described by the Bohr theory (Chapter 2), causing excitation of electrons to higher energy states or producing ionization, with subsequent emission of light. When electrons of inner orbits in heavy elements are displaced, the resultant high energy radiation is classed as X-rays. These rays, which are so useful for internal examination of the human body, are produced by accelerating electrons in a vacuum chamber to energies in the kilovolt range and allowing them to strike a heavy element target. In addition to the X-rays due to transitions in the electronic orbits, a similar radiation called *bremsstrahlung* (German: braking radiation) is produced. It arises from the deflection and resulting acceleration of electrons as they encounter nuclei or atomic electrons.

Beta particles as electrons from nuclear reactions have energies in the range 0.01–1 MeV, and thus are capable of producing large amounts of ionization as they penetrate a substance. As a rough rule of thumb, about 30 eV of energy is required to produce one ion pair. The beta particles lose energy with each event, and eventually are stopped. For electrons of 1 MeV energy, the range, as the typical distance of penetration, is no more than a few millimeters in liquids and solids or a few meters in air.

6.2 HEAVY CHARGED PARTICLE SLOWING BY ATOMS

Charged particles such as protons, alpha particles, or ions such as the fragments of fission are classed as heavy particles, being much more massive than the electron. For the same particle energy they have far less speed than an electron, but they are less readily deflected in their motion than electrons because of their inertia. The mechanism by which heavy ions slow down in matter is primarily electrostatic interaction with atomic electrons. As the positively charged projectile approaches and passes, with the attraction to electrons varying with distance of separation as

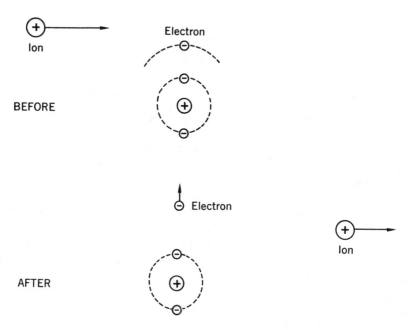

Fig. 6.1.   Interaction of heavy ion with electron.

$1/r^2$, the electron is displaced and gains energy, while the heavy particle loses energy. Figure 6.1 shows conditions before and after the collision schematically. It is found that the kinetic energy lost in one collision is proportional to the square of Z, the number of external electrons in the target atom, and inversely proportional to the kinetic energy of the projectile. A great deal of ionization is produced by the heavy ion as it moves through matter. Although the projectile of energy in the million-electron-volt range loses only a small fraction of its energy in one collision, the amount of energy imparted to the electron can be large compared with its binding to the atom or molecule, and the electron is completely removed. As the result of these interactions, the energy of the heavy ion is reduced and it eventually is stopped in a range that is very much shorter than that for electrons. A 2 MeV alpha particle has a range of about 1 cm in air; it can be stopped by a sheet of paper or the outer layer of skin of the body. Because of these short ranges, there is little difficulty in providing protective shielding against alpha particles.

6.3 HEAVY CHARGED PARTICLE SCATTERING BY NUCLEI

When a high-speed charged ion such as an alpha particle encounters a very heavy charged nucleus, the mutual repulsion of the two particles causes the projectile to move on a hyperbolic path, as in Fig. 6.2. Such a collision can take place in an ionized gas or in a solid if the incoming particle avoids interaction with the external electrons of the atom. The projectile is scattered through an angle that depends on the detailed nature of the collision, i.e., the initial energy and direction of motion of the incoming ion relative to the target nucleus, and the magnitude of electric charge of the interacting particles. Unless the energy of the bombarding particle is very high and it comes within the short range of the nuclear force, there is a small chance that it can enter the nucleus and cause a nuclear reaction.

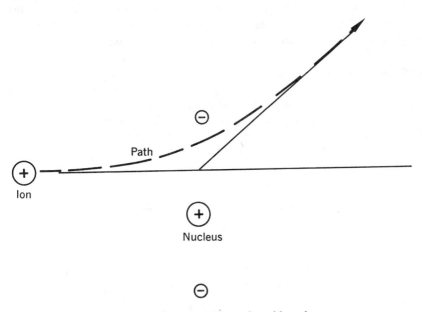

Fig. 6.2.   Interaction of heavy ion with nucleus.

6.4 GAMMA RAY INTERACTIONS WITH MATTER

We now turn to a group of three related processes involving gamma ray photons produced by nuclear reactions. These have energies as high as a few MeV. The interactions include simple scattering of the photon, ionization by it, and a special nuclear reaction known as pair production.

(a) Photon–Electron Scattering

One of the easiest processes to visualize is the interaction of a photon of energy $E = h\nu$ and an electron of rest mass m_0. Although the electrons in a target atom can be regarded as moving and bound to their nucleus, the energies involved are very small (eV) in comparison with those of typical gamma rays (keV or MeV). Thus the electrons may be viewed as free stationary particles. The collision may be treated by the physical principles of energy and momentum conservation. As sketched in Fig. 6.3, the photon is deflected in its direction and loses energy, becoming a photon of new energy $E' = h\nu'$. The electron gains energy and moves away with high speed v and total mass-energy mc^2, leaving the atom ionized. In this *Compton effect*, named after its discoverer, one finds that the greatest photon energy loss occurs when it is scattered backward (180°) from the original direction. Then, if E is much larger than the rest energy of the electron $E_0 = m_0c^2 = 0.51$ MeV, it is found that the *final photon energy E'* is equal to $E_0/2$. On the other hand, if E is much smaller than E_0, the *fractional energy loss* of the photon is $2E/E_0$ (see also Problem 6.3). The derivation of the photon energy loss in general is complicated by the fact that the special theory of relativity must be applied.

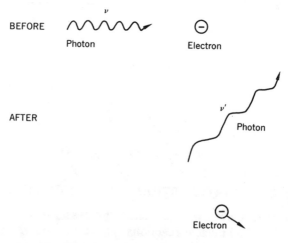

Fig. 6.3.   Photon–electron scattering (Compton effect).

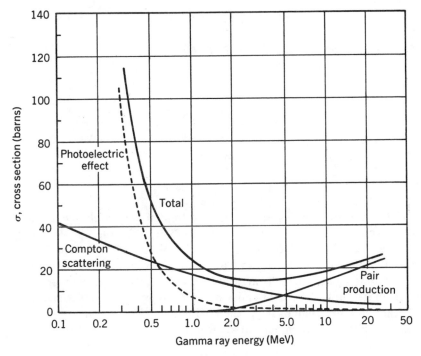

Fig. 6.4. Gamma ray cross sections in lead. (From Raymond L. Murray, *Introduction to Nuclear Engineering*, 2nd Ed. © 1961. Reprinted by permission of Prentice-Hall, Inc., Englewood Cliffs, New Jersey.)

The probability of Compton scattering is expressed by a cross section, which is smaller for larger gamma energies as shown in Fig. 6.4 for the element lead, a common material for shielding against X-rays or gamma rays. We can deduce that the chance of collision increases with each successive loss of energy by the photon, and eventually the photon disappears.

(b) Photoelectric Effect

This process is in competition with scattering. An incident photon of high enough energy dislodges an electron from the atom, leaving a positively charged ion. In so doing, the photon is absorbed and thus lost, see Fig. 6.5. The cross section for the photoelectric effect decreases with increasing photon energy, as sketched in Fig. 6.4 for the element lead.

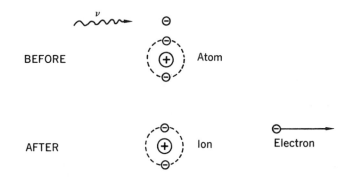

Fig. 6.5. Photoelectric effect.

The above two processes are usually treated separately even though both result in ionization. In the Compton effect, a photon of lower energy survives; but in the photoelectric effect, the photon is eliminated. In each case, the electron released may have enough energy to excite or ionize other atoms by the mechanism described earlier. Also, the ejection of the electron is followed by light emission or X-ray production, depending on whether an outer shell or inner shell is involved. The secondary particles and rays may be more important in the long run than the primary particle.

(c) Electron–Positron Pair Production

The third process to be considered is one in which the photon is converted into matter. This is entirely in accord with Einstein's theory of the equivalence of mass and energy. In the presence of a nucleus, as sketched in Fig. 6.6, a gamma ray photon disappears and two particles appear—an electron and a positron. Since these are of equal charge but of opposite sign, there is no net charge after the reaction, just as before, the gamma ray having zero charge. The law of conservation of charge is thus met. The total new mass produced is twice the mass-energy of the electron, $2(0.51) = 1.02$ MeV, which means that the reaction can occur only if the gamma ray has at least this amount of energy. The cross section for the process of pair production rises from zero as shown in Fig. 6.4 for lead. The reverse process also takes place. As sketched in Fig. 6.7, when an electron and a positron combine, they are annihilated as material particles, and two gamma rays of energy totaling at least 1.02 MeV are released. That there must be two photons is a consequence of the principle of momentum conservation.

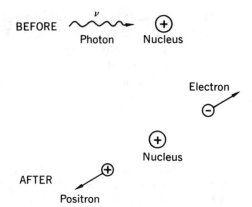

Fig. 6.6.   Pair production.

We note that the total gamma ray cross section curve for a substance, as the sum of the components for Compton effect, photoelectric effect, and pair production, exhibits a minimum around 3 MeV energy. This implies that gamma rays in this vicinity are more penetrating than those of higher or lower energy. In contrast with the case of beta particles and

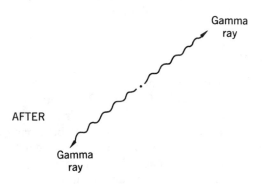

Fig. 6.7.   Pair annihilation.

alpha particles, which have a definite range, a certain fraction of incident gamma rays can pass through any thickness of material. The exponential expression $e^{-\Sigma z}$ as used to describe neutron behavior can be carried over to the attenuation of gamma rays in matter. One can use the mean free path $\lambda = 1/\Sigma$ or better, the half-thickness $0.693/\Sigma$, as the distance in which the intensity of a gamma ray beam is reduced by a factor of two.

6.5 NEUTRON REACTIONS

For completeness, we include a mention of the interaction of neutrons with matter. As discussed in Chapters 4 and 5, neutrons may be scattered by nuclei elastically or inelastically, may be captured with resulting gamma ray emission, or may cause fission. If this energy is high enough, neutrons may induce (n, p) and (n, α) reactions as well.

We are now in a position to understand the connection between neutron reactions and atomic processes. When a high-speed neutron strikes the hydrogen atom in a water molecule, a proton is ejected, resulting in chemical dissociation of the H_2O. A similar effect takes place in molecules of cells in any biological tissue. Now, the proton as a heavy charged particle passes through matter, slowing and creating ionization along its path. Thus two types of radiation damage take place—primary and secondary.

After many collisions, the neutron arrives at a low enough energy that it can be readily absorbed. If it is captured by the proton in a molecule of water or some other hydrocarbon, a gamma ray is released, as discussed in Chapter 4. The resulting deuteron recoils with energy that is much smaller than that of the gamma ray, but still is far greater than the energy of binding of atoms in the water molecule. Again dissociation of the compound takes place, which can be regarded as a form of radiation damage.

6.6 SUMMARY

Radiation of special interest includes electrons, heavy charged particles, photons, and neutrons. Each of the particles tends to lose energy by interaction with the electrons and nuclei of matter, and each creates ionization in different degrees. The ranges of beta particles and alpha particles are short, but gamma rays penetrate in accord with an exponential law. Gamma rays can also produce electron–positron pairs. Neutrons of both high and low energy can create radiation damage in molecular materials.

6.7 PROBLEMS

6.1. The charged particles in a highly ionized electrical discharge in hydrogen gas—protons and electrons, mass ratio $m_p/m_e = 1836$—have the same energies. What is the ratio of the speeds v_p/v_e? Of the momenta p_p/p_e?

6.2. A gamma ray from neutron capture has an energy of 6 MeV. What is its frequency? Its wavelength?

6.3. For 180° scattering of gamma rays or X-rays by electrons, the final energy of the photon is

$$E' = \frac{1}{\dfrac{1}{E} + \dfrac{2}{E_0}}.$$

(a) What is the final photon energy for the 6 MeV gamma ray of Problem 6.2?
(b) Verify that if $E \gg E_0$, then $E' \simeq E_0/2$ and if $E \ll E_0$, $(E - E')/E \simeq 2E/E_0$.
(c) Which approximation should be used for a 6 MeV gamma ray? Verify numerically.

6.4. An electron–positron pair is produced by a gamma ray of 2.26 MeV. What is the kinetic energy imparted to each of the charged particles?

6.5. Estimate the thickness of paper required to stop 2 MeV alpha particles, assuming the paper to be of density 1.29 g/cm³, about the same electronic composition as air, density 1.29×10^{-3} g/cm³.

6.6. The element lead, $M = 206$, has a density of 11.3 g/cm³. Find the number of atoms per cubic centimeter. If the total gamma ray cross section at 3 MeV is 15 barns, what is the macroscopic cross section Σ and the half-thickness $0.693/\Sigma$?

6.7. The range of beta particles of energy greater than 0.8 MeV is given roughly by the relation

$$R(\text{cm}) = \frac{0.55E(\text{MeV}) - 0.16}{\rho(\text{g/cm}^3)}.$$

Find what thickness of aluminum sheet (density 2.7 g/cm³) is enough to stop the betas from phosphorus-32 (see Table 3.1).

7

Fission

Out of the many nuclear reactions known, that resulting in fission has the greatest practical significance. In this chapter we shall describe the mechanism of the process, identify the byproducts, introduce the concept of the chain reaction, and look at the energy yield from the consumption of nuclear fuels.

7.1 THE FISSION PROCESS

The absorption of a neutron by most isotopes involves radiative capture, with the excitation energy appearing as a gamma ray. In certain heavy elements, notably uranium and plutonium, an alternate consequence is observed—the splitting of the nucleus into two massive fragments, a process called fission. Figure 7.1 shows the sequence of events, using the reaction with U-235 to illustrate. In Stage A, the neutron approaches the U-235 nucleus. In Stage B, the U-236 nucleus has been formed, in an excited state. The excess energy in some interactions may be released as a gamma ray, but more frequently, the energy causes distortions of the nucleus into a dumbbell shape, as in Stage C. The parts of the nucleus oscillate in a manner analogous to the motion of a drop of liquid. Because of the dominance of electrostatic repulsion over nuclear attraction, the two parts can separate, as in Stage D. They are then called fission fragments, bearing most of the mass-energy released. They fly apart at high speeds, carrying some 166 MeV of kinetic energy out of the total of around 200 MeV released in the whole process. As the fragments separate, they lose atomic electrons, and the resulting high-speed ions lose

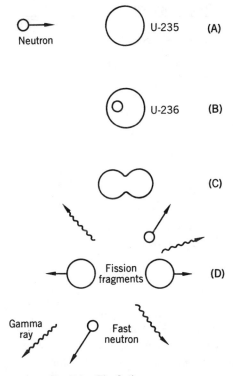

Fig. 7.1. The fission process.

energy by interaction with the atoms and molecules of the surrounding medium. The resultant thermal energy is recoverable if the fission takes place in a nuclear reactor. Also shown in the diagram are the gamma rays and fast neutrons that come off at the time of splitting.

7.2 ENERGY CONSIDERATIONS

The absorption of a neutron by a nucleus such as U-235 gives rise to extra internal energy of the product, because the sum of masses of the two interacting particles is greater than that of a normal U-236 nucleus. We write the first step in the reaction

$$^{235}_{92}U + ^{1}_{0}n \rightarrow (^{236}_{92}U)^*,$$

where the asterisk signifies the excited state. The mass in atomic mass units of (U-236)* is the sum 235.043943 + 1.008665 = 236.052608. However,

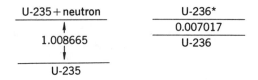

Zero mass level

Fig. 7.2. Excitation energy due to neutron absorption.

U-236 in its ground state has a mass of only 236.045943, lower by 0.007017 amu or 6.5 MeV. This amount of excess energy is sufficient to cause fission. Figure 7.2 shows these energy relations.

The above calculation did not include any kinetic energy brought to the reaction by the neutron, on the grounds that fission can be induced by absorption in U-235 by very slow neutrons. Only one natural isotope, $^{235}_{92}$U, undergoes fission in this way, while $^{239}_{94}$Pu and $^{233}_{92}$U are the main artificial isotopes that do so. Most other heavy isotopes require significantly larger excitation energy to bring the compound nucleus to the required energy level for fission to occur, and the extra energy must be provided by the motion of the incoming neutron. For example, neutrons of at least 0.9 MeV are required to cause fission from U-238, and other isotopes require even higher energy. The precise terminology is as follows: *fissile* materials are those giving rise to fission with slow neutrons; many isotopes are *fissionable*, if enough energy is supplied. It is advantageous to use fast neutrons—of the order of 1 MeV energy—to cause fission. As will be discussed in Chapter 15, the fast reactor permits the "breeding" of nuclear fuel. In a few elements such as californium, spontaneous fission takes place. The isotope $^{252}_{98}$Cf, produced artificially by a sequence of neutron absorptions, has a half-life of 2.65 yr, decaying by alpha emission (97%) and spontaneous fission (3%).

It is not obvious how the introduction of only 6.5 MeV of excitation energy can produce a reaction yielding as much as 200 MeV. The explanation lies in the fact that the sum of the masses of the fission fragments and released neutrons is appreciably smaller than that of the nucleus from which they came (see Problem 7.4).

7.3 BYPRODUCTS OF FISSION

Accompanying the fission process is the release of several neutrons, which are all-important for the practical application to a self-sustaining chain reaction. The number that appear ν (nu) ranges from 1 to 7, with an average in the range 2 to 3 depending on the isotope and the bombarding neutron energy. For example, in U-235 with slow neutrons the average number $\bar{\nu}$ is 2.43. Most of these are released instantly, the so-called *prompt neutrons*, while a small percentage, 0.65% for U-235, appear later as the result of radioactive decay of certain fission fragments. These *delayed neutrons* provide considerable inherent safety and controllability in the operation of nuclear reactors, as we shall see later.

The nuclear reaction equation for fission resulting from neutron absorption in U-235 may be written in general form, letting the chemical symbols for the two fragments be labeled F_1 and F_2 to indicate many possible ways of splitting. Thus

$$^{235}_{92}U + ^{1}_{0}n \rightarrow ^{A_1}_{Z_1}F_1 + ^{A_2}_{Z_2}F_2 + \nu^{1}_{0}n + \text{energy.}$$

The appropriate mass numbers and atomic numbers are attached. One example, in which the fission fragments are isotopes of krypton and barium, is

$$^{235}_{92}U + ^{1}_{0}n \rightarrow ^{90}_{36}Kr + ^{144}_{56}Ba + 2^{1}_{0}n + E.$$

Mass numbers ranging from 75 to 160 are observed, with the most probable at around 92 and 144 as sketched in Fig. 7.3. The ordinate on this graph is the percentage yield of each mass number, e.g., about 6% for mass numbers 90 and 144. If the number of fissions is given, the number of atoms of those types are 0.06 as large.

As a collection of isotopes, these byproducts are called fission products. The isotopes have an excess of neutrons or a deficiency of protons in comparison with naturally occurring elements. For example, the main isotope of barium is $^{137}_{56}Ba$, and a prominent element of mass 144 is $^{144}_{60}Nd$. Thus there are 7 extra neutrons or 4 too few protons in the barium isotope from fission, and it is highly unstable. Radioactive decay, usually involving several emissions of beta particles and delayed gamma

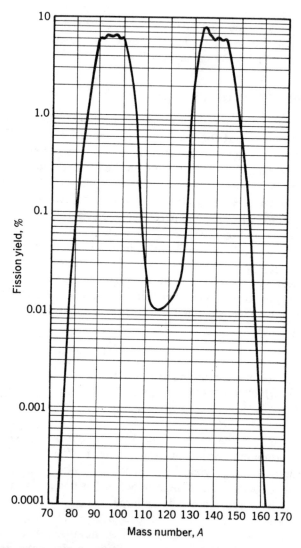

Fig. 7.3. Yield of fission products by mass number. (From Raymond L. Murray, *Introduction to Nuclear Engineering*, 2nd Ed. © 1961. Reprinted by permission of Prentice-Hall, Inc., Englewood Cliffs, New Jersey.)

rays in a chain of events, brings the particles down to stable forms. An example is

$$\,^{90}_{36}\text{Kr} \xrightarrow[33\,\text{sec}]{} \,^{90}_{37}\text{Rb} \xrightarrow[2.91\,\text{min}]{} \,^{90}_{38}\text{Sr} \xrightarrow[27.7\,\text{yr}]{} \,^{90}_{39}\text{Y} \xrightarrow[64\,\text{hr}]{} \,^{90}_{40}\text{Zr}.$$

The hazard associated with the radioactive emanations from fission products is evident when we consider the large yields and the short half-lives.

The total energy release from fission, after all of the particles from decay have been released, is about 200 MeV. This is distributed among the various processes as shown in Table 7.1. The "prompt" gamma rays are emitted as a part of fission, the rest are "decay" gammas. Neutrinos accompany the beta particle emission, but since they are such highly penetrating particles their energy cannot be counted as part of the useful thermal energy yield of the fission process. Thus only about 190 MeV of the fission energy is effectively available. However, several MeV of energy from gamma rays released from nuclei that capture neutrons can also be extracted as useful heat.

Table 7.1.   Energy from Fission, U-235.

	MeV
Fission fragment kinetic energy	166
Neutrons	5
Prompt gamma rays	7
Fission product gamma rays	7
Beta particles	5
Neutrinos	10
Total	200

The average total neutron energy is noted to be 5 MeV. If there are about 2.5 neutrons per fission, the average neutron energy is 2 MeV. When one observes many fission events, the neutrons are found to range in energy from nearly 0 to over 10 MeV, with a most likely value of 0.7 MeV. We note that the neutrons produced by fission are fast, while the cross section for the fission reaction is high for slow neutrons. This fact serves as the basis for the use of a reactor moderator containing a light element that permits neutrons to slow down, by a succession of collisions, to an energy favorable for fission.

Although fission is the dominant process, a certain fraction of the absorptions of neutrons in uranium merely result in radiative capture, according to

$$^{235}_{92}\text{U} + ^1_0\text{n} \rightarrow ^{236}_{92}\text{U} + \gamma.$$

The U-236 is relatively stable, having a half-life of 2.39×10^7 yr. About 15% of the absorptions are of this type, with fission occurring in the remaining 85%. This means that η (eta), the number of neutrons produced per *absorption* in U-235 is lower than ν, the number per *fission*. Thus using $\bar{\nu} = 2.43$, η is $(0.85)(2.43) = 2.07$. The effectiveness of any nuclear fuel is sensitively dependent on the value of η. We find that η is larger for fission induced by fast neutrons than that by slow neutrons.

The possibility of a chain reaction was recognized as soon as it was known that neutrons were released in the fission process. If a neutron is absorbed by the nucleus of one atom of uranium and one neutron is produced, the latter can be absorbed in a second uranium atom, and so on. In order to sustain a chain reaction as in a nuclear reactor or in a nuclear weapon, the value of η must be somewhat above one because of processes that compete with absorption in uranium, such as capture in other materials and escape from the system. The size of η has two important consequences. First, there is a possibility of a growth of neutron population with time. After all extraneous absorption and losses have been accounted for, if one absorption in uranium ultimately gives rise to say 1.1 neutrons, these can be absorbed to give $(1.1)(1.1) = 1.21$, which produce 1.331, etc. The number available increases rapidly with time. Second, there is a possibility of using the extra neutron, over and above the one required to maintain the chain reaction, to produce new fissile materials. "Conversion" involves the production of some new nuclear fuel to replace that used up, while "breeding" is achieved if more fuel is produced than is used.

Out of the hundreds of isotopes found in nature, only one is fissile, $^{235}_{92}\text{U}$. Unfortunately, it is the less abundant of the isotopes of uranium, with weight percentage in natural uranium of only 0.711, in comparison with 99.3% of the heavier isotope $^{238}_{92}\text{U}$. The two other most important fissile materials, plutonium-239 and uranium-233, are "artificial" in the sense that they are man-made by use of neutron irradiation of two *fertile* materials, respectively, uranium-238 and thorium-232. The reactions by which $^{239}_{94}\text{Pu}$ is produced are

$$^{238}_{92}\text{U} + ^1_0\text{n} \rightarrow ^{239}_{92}\text{U},$$

$$^{239}_{92}\text{U} \xrightarrow[2.35\,\text{min}]{} {}^{239}_{93}\text{Np} + ^{\ 0}_{-1}\text{e},$$

$$^{239}_{93}\text{Np} \xrightarrow[2.35\,\text{day}]{} {}^{239}_{94}\text{Pu} + ^{\ 0}_{-1}\text{e},$$

while those yielding $^{233}_{92}U$ are

$$^{232}_{90}Th + ^1_0n \rightarrow ^{233}_{90}Th,$$

$$^{233}_{90}Th \xrightarrow[22\ min]{} {}^{233}_{91}Pa + {}^0_{-1}e,$$

$$^{233}_{91}Pa \xrightarrow[27\ day]{} {}^{233}_{92}U + {}^0_{-1}e.$$

The half-lives for decay of the intermediate isotopes are short compared with times involved in the production of these fissile materials; and for many purposes, these decay steps can be ignored. It is important to note that although uranium-238 is not fissile, it can be put to good use as a fertile material for the production of plutonium-239, so long as there are enough free neutrons available.

7.4 ENERGY FROM NUCLEAR FUELS

The practical significance of the fission process is revealed by calculation of the amount of uranium that is consumed to obtain a given amount of energy. Each fission yields 190 MeV of useful energy, which is also $(190\ MeV)(1.60 \times 10^{-13}\ J/MeV) = 3.04 \times 10^{-11}\ J$. Thus the number of fissions required to obtain 1 W-sec of energy is $1/(3.04 \times 10^{-11}) = 3.3 \times 10^{10}$. The number of U-235 atoms consumed in a thermal reactor is larger by the factor $1/0.85 = 1.18$ because of the formation of U-236 in part of the reactions.

In one day's operation of a reactor per megawatt of thermal power, the number of U-235 nuclei burned is

$$\frac{(10^6\ W)(3.3 \times 10^{10}\ fissions/W\text{-}sec)(86,400\ sec/day)}{0.85\ fissions/absorptions}$$

$$= 3.35 \times 10^{21}\ absorptions/day.$$

Then since 235 g corresponds to Avogadro's number of atoms 6.02×10^{23}, the U-235 weight consumed at 1 MW power is

$$\frac{(3.35 \times 10^{21}\ day^{-1})(235\ g)}{(6.02 \times 10^{23})} \simeq 1.3\ g/day.$$

In other words, 1.3 g of fuel is used per megawatt-day of useful thermal energy released. In a typical reactor, which produces 3000 MW of thermal power, the U-235 fuel consumption is about 4 kg/day. To produce the same energy by use of fossil fuels such as coal, oil, or gas, millions of times as much weight would be required.

7.5 SUMMARY

Neutron absorption by the nuclei of heavy elements gives rise to fission, in which heavy fragments, fast neutrons, and other radiations are released. Fissile materials are natural U-235 and man-made isotopes Pu-239 and U-233. Many different radioactive isotopes are released in the fission process, and more neutrons are produced than are used, which makes possible a chain reaction and under certain conditions "conversion" and "breeding" of new fuels. Useful energy amounts to 190 MeV per fission, requiring only 1.3 g of U-235 to be consumed to obtain 1 MW-day of energy.

7.6 PROBLEMS

7.1. Calculate the excitation energy in (Pu-240)* resulting from the absorption of a neutron of mass 1.00866 in Pu-239, mass 239.05217, assuming the mass of Pu-240 to be 240.05384.

7.2. If three neutrons and a xenon-133 atom ($^{133}_{54}$Xe) are produced when a U-235 atom is bombarded by a neutron, what is the second fission product isotope?

7.3. Neglecting neutron energy and momentum effects, what are the energies of the two fission fragments if their mass ratio is $3:2$?

7.4. Calculate the energy yield from the reaction

$$^{235}_{92}U + ^1_0n \rightarrow ^{140}_{55}Cs + ^{92}_{37}Rb + 4\,^1_0n + E$$

using atomic masses 139.91711 for cesium and 91.91914 for rubidium.

7.5. The value of η for U-233 for thermal neutrons is approximately 2.29. Using the cross sections for capture $\sigma_c = 48$ barns and fission $\sigma_f = 531$ barns, deduce the value of ν, the number of neutrons per fission.

7.6. A mass of 8000 kg of slightly enriched uranium (2% U-235, 98% U-238) is exposed for 30 days in a reactor operating at heat power 2000 MW. Neglecting consumption of U-238, what is the final fuel composition?

7.7. The per capita consumption of electrical energy in the United States is about 50 kWh/day. If this were provided by fission with $\frac{2}{3}$ of the heat wasted, how much U-235 would each person use per day?

8

Fusion

When two light nuclear particles combine or "fuse" together, energy is released because the product nuclei have less mass than the original particles. Such fusion reactions can be caused by bombarding targets with charged particles, using an accelerator, or by raising the temperature of a gas to high enough level that nuclear reactions take place. In this chapter we shall describe the interactions in the microscopic sense and discuss the phenomena that affect our ability to achieve a practical large-scale source of energy from fusion.

8.1 FUSION REACTIONS

The possibility of release of large amounts of nuclear energy can be seen by comparing the masses of nuclei of low atomic number. Suppose that one could combine two hydrogen nuclei and two neutrons to form the helium nucleus. In the reaction

$$2\,^1_1\text{H} + 2\,^1_0\text{n} \rightarrow\,^4_2\text{He},$$

the mass-energy difference (using atom masses) is

$$2(1.007825) + 2(1.008665) - 4.002603 = 0.030377 \text{ amu},$$

which corresponds to 28.2 MeV energy. A comparable amount of energy would be obtained by combining four hydrogen nuclei to form helium plus two positrons

$$4\,^1_1\text{H} \rightarrow\,^4_2\text{He} + 2\,^0_1\text{e}.$$

This reaction in effect takes place in the sun and other stars through the so-called carbon cycle, a complicated chain of events involving hydrogen and isotopes of the elements carbon, oxygen, and nitrogen. The cycle is extremely slow, however, and is not suitable for terrestrial application. The sun's large energy yield is due to its tremendous mass of materials, not to the large rate of nuclear reactions per unit volume.

In the "hydrogen bomb," on the other hand, the high temperatures created by a fission reaction cause the fusion reaction to proceed in a rapid and uncontrolled manner. Between these extremes is the possibility of achieving a controlled fusion reaction that utilizes inexpensive and abundant fuels. As yet, a practical fusion device has not been developed, and considerable research and development will be required to reach that goal. Let us now examine the nuclear reactions that might be employed. There appears to be no mechanism by which four separate nuclei can be made to fuse directly, and thus combinations of two particles must be sought.

The most promising reactions make use of the isotope deuterium. It is present in hydrogen as in water with abundance only 0.015%, i.e., there is one atom of $_1^2H$ for every 6700 atoms of $_1^1H$, but since our planet has enormous amounts of water, the fuel available is almost inexhaustible. Four reactions are important:

$$_1^2H + _1^2H \rightarrow _1^3H + _1^1H + 4.03 \text{ MeV},$$

$$_1^2H + _1^2H \rightarrow _2^3He + _0^1n + 3.27 \text{ MeV},$$

$$_1^2H + _1^3H \rightarrow _2^4He + _0^1n + 17.6 \text{ MeV},$$

$$_1^2H + _2^3He \rightarrow _2^4He + _1^1H + 18.3 \text{ MeV}.$$

The fusion of two deuterons—deuterium nuclei—in what is designated the D–D reaction results in two processes of equal likelihood. The other reactions yield more energy but involve the artificial isotopes tritium, $_1^3H$, and helium-3. We note that the products of the first and second equations appear as reactants in the third and fourth equations. This suggests that a composite process might be feasible. Suppose that each of the reactions could be made to proceed at the same rate, along with twice the reaction of neutron capture in hydrogen

$$_1^1H + _0^1n \rightarrow _1^2H + 2.2 \text{ MeV}.$$

Adding all of the equations, we find that the net effect is to convert deuterium into helium according to

$$4\,{}^{2}_{1}\text{H} \rightarrow 2\,{}^{4}_{2}\text{He} + 47.7\,\text{MeV}.$$

The energy yield per atomic mass unit of deuterium fuel would thus be about 6 MeV, which is much more favorable than the yield per atomic mass unit of U-235 burned, which is only $190/235 = 0.85$ MeV.

8.2 ELECTROSTATIC AND NUCLEAR FORCES

The reactions described above do not take place merely by mixing the ingredients because of the very strong force of electrostatic repulsion between the charged nuclei. Only by giving one or both of the particles a high speed can they be brought close enough to each other for the strong nuclear force to dominate the electrical force. This behavior is in sharp contrast with the ease with which neutrons interact with nuclei.

There are two consequences of the fact that the coulomb force between two charges of atomic numbers Z_1 and Z_2 varies with separation R according to $Z_1 Z_2 / R^2$. First, we see that fusion is unlikely in elements other than those low in the periodic table. Second, the force and corresponding potential energy of repulsion is very large at the 10^{-15} m range of nuclear forces, and thus the chance of reaction is negligible unless particle energies are of the order of kilovolts. Figure 8.1 shows the cross section for the D–D reaction. The strong dependence on energy is noted, with σ_{DD} rising by a factor of 1000 in the range 10–75 keV.

Energies in the kilo-electron-volt and million-electron-volt range can be achieved by a variety of charged particle accelerators. Bombardment of a solid or gaseous deuterium target by high-speed deuterons gives fusion reactions, but most of the particle energy goes into electrostatic interactions that merely heat up the bulk of the target. The amount of energy required to operate the accelerator greatly exceeds the recoverable fusion energy, and thus some other technique is required.

8.3 THERMONUCLEAR REACTIONS IN IONIC PLASMA

The most promising medium in which to obtain the high particle energies that are needed for practical fusion is the plasma. It consists of a completely ionized gas as in an electrical discharge created by the acceleration of electrons. Equal numbers of electrons and deuterons are present, making the medium electrically neutral. Through the injection of enough energy into the plasma its temperature can be increased, and the deuterons reach the speed for fusion to be favorable. The term

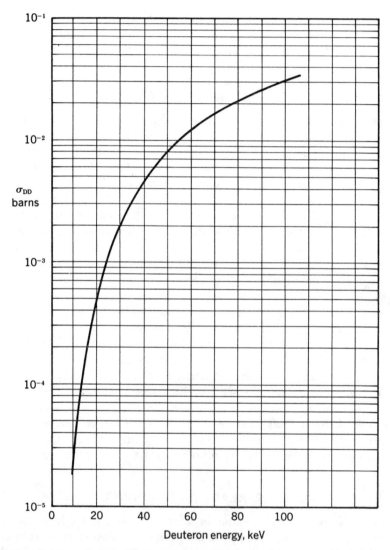

Fig. 8.1. Cross section for D–D reaction. (From Raymond L. Murray, *Introduction to Nuclear Engineering*, 2nd Ed. © 1961. Reprinted by permission of Prentice-Hall, Inc., Englewood Cliffs, New Jersey.)

thermonuclear is applied to reactions induced by high thermal energy, and the particles obey a speed distribution similar to that of a gas, as discussed in Chapter 2.

The temperatures to which the plasma must be raised are extremely high, as we can see by expressing an average particle energy in terms of temperature, using the kinetic relation

$$\bar{E} = \frac{3}{2} kT.$$

For example, even if $\bar{E}$ is as low as 10 keV, the temperature is

$$T = \frac{2}{3} \frac{(10^4 \, \text{eV})(1.60 \times 10^{-19} \, \text{J/eV})}{1.38 \times 10^{-23} \, \text{J/°K}}$$

or

$$T = 77,000,000°\text{K}.$$

Such a temperature greatly exceeds the temperature of the surface of the sun, and is far beyond any temperature at which ordinary materials melt and vaporize. The plasma must be created and heated to the necessary temperature under the constraint of electric and magnetic fields. Such forces on the plasma are required to assure that thermal energy is not prematurely lost. Moreover, the plasma must remain intact long enough for many nuclear reactions to occur, which is difficult because of inherent instabilities of such highly charged media.

The achievement of a practical energy source is further limited by the phenomenon of radiation losses. In Chapter 6 we discussed the bremsstrahlung radiation produced when electrons experience acceleration. Conditions are ideal for the generation of such electromagnetic radiation since the high-speed electrons in the plasma at elevated temperature experience continuous accelerations and decelerations as they interact with other charges. The radiation can readily escape from the region, because the number of target particles is very small. In a typical plasma, the number density of electrons and deuterons is 10^{15}, which corresponds to a rarefied gas. The amount of radiation production (and loss) increases with temperature at a slower rate than does the energy released by fusion, as shown in Fig. 8.2. At what is called the "ignition temperature," the lines cross. Only for temperatures above that value, 400,000,000°K in the case of the D–D reaction, will there be a net energy yield, assuming that the radiation is lost. In a later chapter we shall describe some of the devices that have been used to explore the possibility of achieving a fusion reactor.

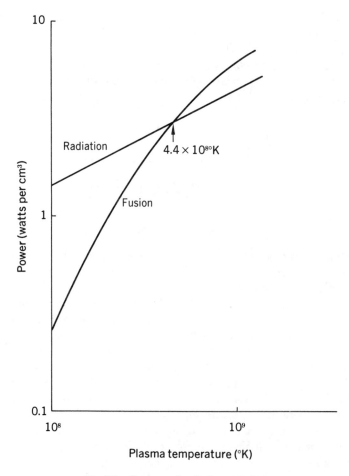

Fig. 8.2.   Fusion and radiation energies.

8.4 SUMMARY

Nuclear energy is released when nuclei of two light elements combine. The most favorable fusion reactions involve deuterium, which is a natural component of water and thus is a very abundant fuel. The reaction takes place only when the particles have a high enough speed to overcome the electrostatic repulsion of their charges. In a highly ionized electrical medium, the plasma, at temperatures of the order of 400,000,000°K, the fusion energy can exceed the energy loss due to radiation.

8.5 PROBLEMS

8.1. Calculate the energy release in amu and MeV from the combination of four hydrogen atoms to form a helium atom and two positrons (each of mass 0.000549 amu).

8.2. Verify the energy yield for the reaction $^2_1H + {}^3_2He \rightarrow {}^4_2He + {}^1_1H + 18.3\,MeV$, noting atomic masses (in order) 2.01410, 3.01603, 4.00260, and 1.00782.

8.3. To obtain 3000 MW of power from a fusion reactor, in which the effective reaction is $2\,{}^2_1H \rightarrow {}^4_2He + 23.8\,MeV$, how many grams per day of deuterium would be needed? If all of the ^2_1H could be extracted from water, how many kilograms of water would have to be processed per day?

8.4. The reaction rate relation $nvN\sigma$ can be used to estimate the power density of a fusion plasma. (a) Find the speed v_D of 100 keV deuterons. (b) Assuming that deuterons serve as both target and projectile, such that the effective v is $v_D/2$, find what particle number density would be needed to achieve a power density of $1\,kW/cm^3$.

Part II Nuclear Systems

The atomic and nuclear concepts we have described provide the basis for the operation of a number of devices, machines, or processes, ranging from very small radiation detectors to mammoth plants to process uranium or to generate electrical power. These systems may be designed to produce nuclear energy, or to make practical use of it, or to apply byproducts of nuclear reactions for beneficial purposes. In the next several chapters we shall explain the construction and operating principles of nuclear systems, referring back to basic concepts and looking forward to appreciating their impact on human affairs.

9

Particle Accelerators

A device that provides forces on charged particles by some combination of electric and magnetic fields and brings the ions to high speed and kinetic energy is called an accelerator. Many types have been developed for the study of nuclear reactions and basic nuclear structure, with an ever-increasing demand for higher particle energy. In this chapter we shall review the nature of the forces on charges and describe the arrangement and principle of operation of several important kinds of particle accelerators.

9.1 ELECTRIC AND MAGNETIC FORCES

Let us recall how charged particles are influenced by electric and magnetic fields. First, visualize a pair of parallel metal plates separated by a distance d as in the sample capacitor shown in Fig. 9.1. A potential difference V and electric field $\mathscr{E} = V/d$ are provided to the region of low gas pressure by a direct-current voltage supply such as a battery. If an electron of mass m and charge e is released at the negative plate, it will experience a force $\mathscr{E}e$, and its acceleration will be $\mathscr{E}e/m$. It will gain speed, and on reaching the positive plate it will have reached a kinetic energy $\frac{1}{2}mv^2 = Ve$. Thus its speed is $v = \sqrt{2Ve/m}$. For example, if V is 100 volts, the speed of an electron ($m = 9.1 \times 10^{-31}$ kg and $e = 1.60 \times 10^{-19}$ coulombs) is easily found to be 5.9×10^6 m/sec.

Next, let us introduce a charged particle of mass m, charge e, and speed v into a region with uniform magnetic field B, as in Fig. 9.2. If the charge enters in the direction of the field lines, it will not be affected, but if it

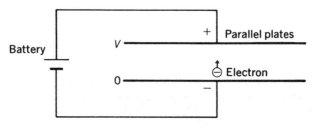

Fig. 9.1. Capacitor as accelerator.

enters perpendicularly to the field, it will move at constant speed on a circle. Its radius, called the radius of gyration, is $r = mv/eB$, such that the stronger the field or the lower the speed, the smaller will be the radius of motion. Letting the angular speed be $w = v/r$ we see that $w = eB/m$. If the charge enters at some other angle, it will move in a path called a helix, like a wire door spring.

Finally, let us release a charge in a region where the magnetic field B is changing with time. If the electron were inside the metal of a circular loop of wire of area A as in Fig. 9.3, it would experience an electric force induced by the change in magnetic flux BA. The same effect would take place without the presence of the wire, of course.

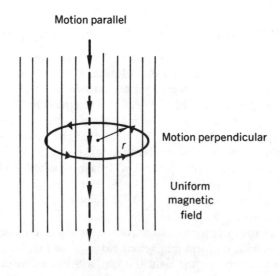

Fig. 9.2. Electric charge motion in uniform magnetic field.

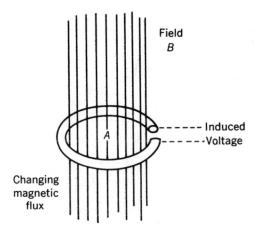

Field
B

---- Induced
--- -Voltage

A

Changing
magnetic
flux

Fig. 9.3.   Magnetic induction.

9.2   HIGH-VOLTAGE MACHINES

One way to accelerate ions to high speed is to provide a large potential difference between a source of charges and a target. In effect, the phenomenon of lightning, in which a discharge from charged clouds to the earth takes place, is produced in the laboratory. Two devices of this type are commonly used. The first is the voltage multiplier or Cockroft–Walton machine, Fig. 9.4, which has a circuit that charges capacitors in parallel and discharges them in series.

The second is the electrostatic generator or Van de Graaff accelerator, the principle of which is sketched in Fig. 9.5. An insulated metal shell is raised to high potential by bringing it charge on a moving belt, permitting the acceleration of positive charges such as protons or deuterons. Particle energies of the order of 5 MeV are possible, with a very small spread in energy.

9.3   LINEAR ACCELERATOR

Rather than giving a charge one large acceleration with a high voltage, it can be brought to high speed by a succession of accelerations through relatively small potential differences, as in the linear accelerator, sketched in Fig. 9.6. It consists of a series of accelerating electrodes in the form of tubes with alternating electric potentials applied as shown. An electron or ion gains energy in the gaps between tubes and "drifts" without change of

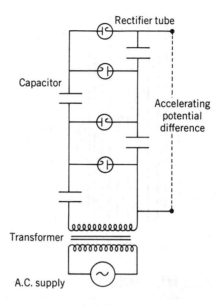

Fig. 9.4. Cockroft–Walton circuit.

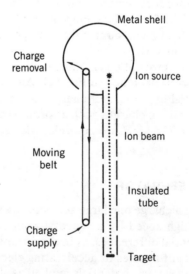

Fig. 9.5. Van de Graaff accelerator.

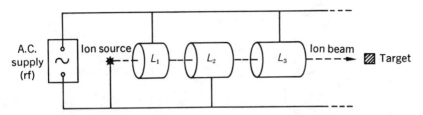

Fig. 9.6. Simple linear accelerator. (From Raymond L. Murray and Grover C. Cobb, *PHYSICS: Concepts and Consequences*, © 1970. Reprinted by permission of Prentice-Hall, Inc., Englewood Cliffs, New Jersey.)

energy while inside the tube, where the field is nearly zero. By the time the charge reaches the next gap, the voltage is again correct for acceleration. Because the ion is gaining speed along the path down the row of tubes, their lengths l must be successively longer in order for the time of flight in each to be constant. The time to go a distance l is l/v, which is equal to the half-period of the voltage cycle $T/2$. Particle energies of 20 billion electron-volts are obtained in the two-mile-long Stanford accelerator.

9.4 CYCLOTRON

Successive electrical accelerations by electrodes and circular motion within a magnetic field are combined in the cyclotron. As sketched in Fig. 9.7, ions such as protons, deuterons, or alpha particles are provided by a source at the center of a vacuum chamber located between the poles of a large electromagnet. Two hollow metal boxes called "dees" (in the shape of the letter D) are supplied with alternating voltages in correct frequency and opposite polarity. In the gap between dees, an ion gains energy as in the linear accelerator, then moves on a circle while inside the field-free region, guided by the magnetic field. Each crossing of the gap with potential difference V gives impetus to the ion with an energy gain Ve, and the radius of motion increases according to $r = v/w$, where $w = eB/m$ is the angular speed. The unique feature of the cyclotron is that the time required for one complete revolution, $T = 2\pi/w$, is independent of the radius of motion of the ion. Thus it is possible to use a synchronized alternating potential of constant frequency v, angular frequency $w = 2\pi v$, to provide acceleration at the right instant.

For example, in a magnetic field B of 0.5 Wb/m^2 the angular speed for

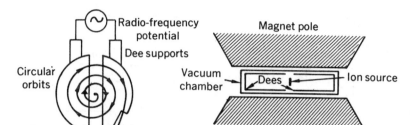

Fig. 9.7. Cyclotron. (From Raymond L. Murray and Grover C. Cobb, *PHYSICS: Concepts and Consequences*, © 1970. Reprinted by permission of Prentice-Hall, Inc., Englewood Cliffs, New Jersey.)

deuterons of mass 3.3×10^{-27} kg and charge 1.6×10^{-19} coulombs is

$$w = \frac{eB}{m} = \frac{(1.6 \times 10^{-19})(0.5)}{3.3 \times 10^{-27}} = 2.4 \times 10^{7}/\text{sec}.$$

Equating this to the angular frequency for the power supply, $w = 2\pi\nu$, we find $\nu = (2.4 \times 10^{7})/2\pi = 3.8 \times 10^{6} \text{ sec}^{-1}$, which is in the radio-frequency range.

The path of ions is approximately a spiral. When the outermost radius is reached and the ions have full energy, a beam is extracted from the dees by special electric and magnetic fields, and allowed to strike a target, in which nuclear reactions take place.

9.5 BETATRON

Electrons are brought to high speeds in the induction accelerator or betatron. A changing magnetic flux provides an electric field and a force on the charges, while they are guided in a path of constant radius. Figure 9.8 shows the vacuum chamber in the form of a doughnut placed between specially shaped magnetic poles. The force on electrons of charge e is in the direction tangential to the orbit of radius r. The rate at which the average magnetic field within the loop changes is $\Delta B/\Delta t$, provided by varying the current in the coils of the electromagnet. The amount of force is†

---

†To show this, note that the area within the circular path is $A = \pi r^{2}$ and the magnetic flux is $\Phi = BA$. According to Faraday's law of induction, if the flux changes by $\Delta\Phi$ in a time Δt, a potential difference around a circuit of $V = \Delta\Phi/\Delta t$ is produced. The corresponding electric field is $\mathscr{E} = V/2\pi r$, and the force is $e\mathscr{E}$. Combining, the relation quoted is obtained.

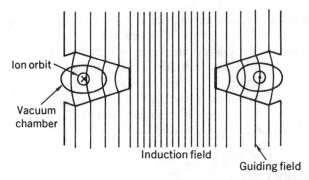

Fig. 9.8. Betatron. (From Raymond L. Murray and Grover C. Cobb, *PHYSICS: Concepts and Consequences*, © 1970. Reprinted by permission of Prentice-Hall, Inc., Englewood Cliffs, New Jersey.)

$$F = \frac{er}{2}\frac{\Delta B}{\Delta t}.$$

The charge continues to gain energy while remaining at the same radius if the magnetic field there is half the average field within the loop. The acceleration to energies in the million-electron-volt range takes place in the fraction of a second that it takes for the alternating magnetic current to go through a quarter-cycle.

The speeds reached in a betatron are high enough to require the use of relativistic formulas (Chapter 1). Let us find the mass m and speed v for an electron of kinetic energy $E_k = 1$ MeV. Rearranging the equation for kinetic energy, the ratio of m to the rest mass m_0 is

$$\frac{m}{m_0} = 1 + \frac{E_k}{m_0 c^2}.$$

Recalling that the rest energy $E_0 = mc^2$ for an electron is 0.51 MeV we obtain $m/m_0 = 1 + 1/0.51 = 2.96$. Solving Einstein's equation $m/m_0 = 1/\sqrt{1-(v/c)^2}$ for the speed, we find that $v = c\sqrt{1-(m_0/m)^2} = 0.94c$. Thus the 1 MeV electron's speed is close to that of light, $c = 3.0 \times 10^8$ m/sec, i.e., $v = 2.8 \times 10^8$ m/sec. If instead we impart a kinetic energy of 100 MeV to an electron, its mass increases by a factor 297 and its speed becomes $0.999995c$.

One of the most powerful accelerators in the world is located at the National Accelerator Laboratory near Chicago. By a combination of accelerating devices of the type described in this chapter, it gives protons an energy of around 400 billion-electron-volts. First, a Cockroft–Walton machine provides particle energies of 0.75 MeV. Then, the ions are raised to an energy of 200 MeV by use of a linear accelerator, and injected for final acceleration into a synchrotron, which involves both changing magnetic fields and radio-frequency electric fields.

By the use of accelerators of greater sophistication and higher particle energy many new subnuclear particles such as mesons and xi, sigma, and lambda particles have been discovered, and the internal structure of nuclei has become better understood.

9.6 SUMMARY

Charged particles such as electrons and ions of light elements are brought to high speed and energy by particle accelerators, which employ electric and magnetic fields in various ways. In the high-voltage machines a beam of ions is accelerated directly through a large potential difference, produced by special voltage multiplier circuits or by carrying charge to a positive electrode; in the linear accelerator, ions are given successive accelerations in gaps between tubes lined up in a row; in the cyclotron, the ions are similarly accelerated but move in circular orbits because of the applied magnetic field; in the betatron, a changing magnetic field produces an electric field that accelerates electrons to relativistic speeds. High-energy nuclear physics research is carried out through the use of such accelerators.

9.7 PROBLEMS

9.1. Calculate the potential difference required to accelerate an electron to speed 2×10^5 m/sec.

9.2. What is the proper frequency for a voltage supply to a linear accelerator tube of length 0.6 m in order to accelerate protons to speed 3×10^6 m/sec?

9.3. Find the time for one revolution of a deuteron in a uniform magnetic field of 1 Wb/m^2.

9.4. Develop a working formula for the final energy of cyclotron ions of mass m, charge q, exit radius R, in a magnetic field B. (Use nonrelativistic energy relations.) Answer: $E = (qBR)^2/2m$.

9.5. What magnetic field strength (Wb/m^2) is required to accelerate deuterons in a cyclotron of radius 2.5 m to energy 5 MeV?

9.6. What is the factor by which the mass is increased and what fraction of the speed of light do protons of 200 billion-electron-volts have?

10

Isotope Separators

All of our technology is based on materials in various forms—compounds, alloys, and mixtures. Examples that immediately come to mind are copper for conduction of electricity, steel and concrete for building construction, drugs for medical treatment, and gasoline for propulsion of automobiles. Materials that are mixtures may be prepared by heating and mechanical action; those that are pure elements or compounds may be produced by chemical processing. For several materials used in the nuclear field, however, individual isotopes of elements or a specified combination of isotopes are required. Two important examples are $^{235}_{92}U$ and $^{2}_{1}H$. Since isotopes of a given element have the same atomic number Z, they are essentially identical chemically and thus a physical method that distinguishes between particles on the basis of mass number A is required. In this chapter we shall describe three devices by which uranium isotopes are separated. One is based on differences in ion motion in a magnetic field; the other two on differences in the diffusion of particles through a membrane or against a centrifugal force. Calculations on the amounts of material that must be processed to obtain nuclear fuel will be presented, and estimates of costs given.

10.1 MASS SPECTROGRAPH

We recall from Chapter 9 that a particle of mass m, charge q, and speed v will move in a circular path of radius r if injected perpendicular to a magnetic field of strength B, according to the relation $r = mv/qB$. In the mass spectrograph (Fig. 10.1), ions of the element whose isotopes are to

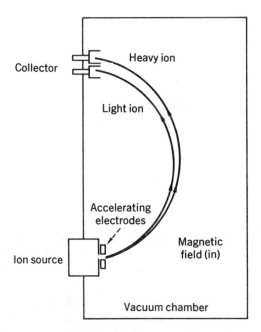

Fig. 10.1.   Mass spectrograph.

be separated are produced in an electrical discharge and accelerated through a potential difference V to provide a kinetic energy $\frac{1}{2}mv^2 = qV$. The charges move freely in a chamber maintained at very low gas pressure, guided in semicircular paths by the magnetic field. The heavier ions have a larger radius of motion than the light ions, and the two may be collected separately. It is found (see Problem 10.1) that the distance between the points at which ions are collected is proportional to the difference in the square roots of the masses. The spectrograph can be used to measure masses with some accuracy, or to determine the relative abundance of isotopes in a sample, or to enrich an element in a certain desired isotope.

The electromagnetic process is especially useful for separating light isotopes and those for which small quantities are needed. However, since a large amount of electrical energy is required to provide the magnetic field and the ion acceleration, the cost of large-scale uranium isotope separation by the method is prohibitive, and an alternate process, gaseous diffusion, is the principal one employed to provide reactor fuels.

10.2 GASEOUS DIFFUSION SEPARATOR

The principle of this process can be illustrated by a simple experiment, Fig. 10.2. A container is divided into two parts by a porous membrane and air is introduced on both sides. Recall that air is a mixture of 80% nitrogen, $A = 14$, and 20% oxygen, $A = 16$. If the pressure on one side is raised, the relative proportion of nitrogen on the other side increases. The separation effect can be explained on the basis of particle speeds. The average kinetic energies of the heavy (H) and light (L) molecules in the gas mixture are the same, $E_H = E_L$, but since the masses are different, the typical particle speeds bear a ratio

$$\frac{v_L}{v_H} = \sqrt{\frac{m_H}{m_L}}$$

Now the number of molecules of a given type that hit the membrane each second is proportional to nv, in analogy to neutron motion discussed in Chapter 5. Those with higher speed thus have a higher probability of passing through the holes in the porous membrane, called "barrier."

The physical arrangement of one processing unit of a gaseous diffusion plant for the separation of uranium isotopes U-235 and U-238 is shown in

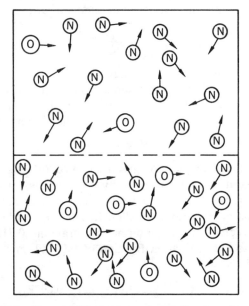

Fig. 10.2. Gaseous diffusion separation of nitrogen and oxygen.

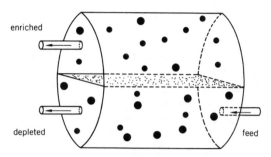

Fig. 10.3. Gaseous diffusion stage.

Fig. 10.3. A thin nickel alloy serves as the barrier material. In this "stage," gas in the form of the compound uranium hexafluoride (UF$_6$) is pumped in as feed and removed as two streams. One is enriched and one depleted in the compound ^{235}UF$_6$, with corresponding changes in ^{238}UF$_6$. Because of the very small mass difference of particles of molecular weight 349 and 352 the amount of separation is small and many stages in series are required in what is called a cascade.

Any isotope separation process causes a change in the relative numbers of molecules of the two species. Let n_H and n_L be the number of molecules in a sample of gas. Their *abundance ratio* is defined as

$$R = \frac{n_L}{n_H}.$$

For example, in ordinary air $R = 80/20 = 4$.

The effectiveness of an isotope separation process is dependent on a quantity called the separation factor r. If we supply gas at one abundance ratio R, the ratio R' on the low-pressure side of the barrier is given by

$$R' = rR.$$

If only a very small amount of gas is allowed to diffuse through the barrier, the separation factor is given by $r = \sqrt{m_H/m_L}$, which for UF$_6$ is 1.0043. However, for a more practical case, in which half the gas goes through, the separation factor is smaller, 1.0030 (see Problem 10.2). Let us calculate the effect of one stage on natural uranium, 0.711% by weight, corresponding to a U-235 atom fraction of 0.00720, and an abundance ratio of 0.00725. Now

$$R' = rR = (1.0030)(0.00725) = 0.00727.$$

The amount of enrichment is very small. By processing the gas in a series

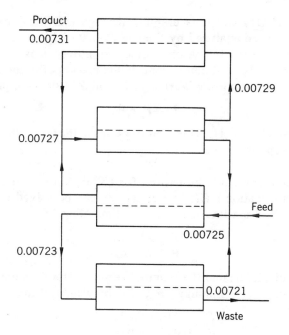

Fig. 10.4.   Gaseous diffusion cascade.

of s stages, each one of which provides a factor r, the abundance ratio is increased by a factor r^s. If R_f and R_p refer to feed and product, respectively, $R_p = r^s R_f$. For $r = 1.0030$ we can easily show that 2375 enriching stages are needed to go from $R_f = 0.00725$ to highly enriched 90% U-235, i.e., $R_p = 0.9/(1 - 0.9) = 9$. Figure 10.4 shows the arrangement of several stages in an elementary cascade, and indicates the value of R at various points. The feed is natural uranium, the product is enriched in U-235, and the waste is depleted in U-235.

10.3   URANIUM ENRICHMENT COSTS

A gaseous diffusion plant is very expensive, of the order of a billion dollars, because of the size and number of components such as separators, pumps, valves, and controls, but the process is basically simple. The plant runs continuously with few operating personnel. The principal operating cost is for the electrical power to provide the pressure differences and perform work on the gas.

The flow of UF_6 and thus uranium through individual stages or the whole plant can be analyzed by the use of material balances. Since the plant operates continuously, one could use atomic mass units but the kilogram per day is most convenient. If the masses of uranium that flow per day through a plant are labeled F, P, and W (for feed, product, and waste), then

$$F = P + W.$$

Letting x stand for the U-235 weight fractions in the flows, the balance for the light isotope is

$$x_f F = x_p P + x_w W.$$

(A similar equation could be written for U-238, but it would contain no additional information.) The two equations can be solved to obtain the ratio of feed and product mass rates. Eliminating W,

$$\frac{F}{P} = \frac{x_p - x_w}{x_f - x_w}.$$

For example, let us find the required feed of natural uranium to obtain 1 kg/day of product containing 3% U-235 by weight, if the waste is 0.2% U-235. Now

$$\frac{F}{P} = \frac{0.03 - 0.002}{0.00711 - 0.002} = 5.5$$

and thus the feed is 5.5 kg/day. We note that W is 4.5 kg/day, which shows that large amounts of depleted uranium "tails" must be stored for each kilogram of U-235 produced. The U-235 content of the tails is too low for use in conventional reactors, but the breeder reactor can convert the U-238 into plutonium, as will be discussed in Chapter 15.

As expected, the higher is the enrichment, the greater is the cost of the uranium product. Table 10.1, columns 1 and 2, shows that the cost if purchased outright varies by a factor of 5150 over the range of weight percents. The cost of the U-235 goes from $1250 to $14,300 per kilogram. Column 3 shows the ratio feed/product, where the figure 5.479 for 3% corresponds to our calculated 5.5. Column 4 is used to calculate the cost of performing enrichment on uranium supplied by a customer. The "separative work" is proportional to the energy required for each kilogram of uranium product handled, and a cost of each separative work unit (SWU) is set by the United States Atomic Energy Commission on the basis of current expenses of production.

Suppose that a utility company wants uranium at 3% enrichment for use in its nuclear reactor, and supplies the necessary natural uranium. By use of Table 10.1, the feed required is 5.479 kg for each kilogram of

Table 10.1.   Nuclear Fuel and Enrichment Costs.†

Weight percent U-235	Cost of enriched U ($/kg)	Feed/product	Separative work (SWU)
0.2	2.50	—	—
0.5	7.11	0.587	−0.173
0.711	23.46	1.000	0
0.8	31.55	1.174	0.104
1.0	51.37	1.566	0.380
2.0	167.12	3.523	2.194
3.0	294.32	5.479	4.306
5.0	561.12	9.393	8.851
10.0	1253.14	19.178	20.863
90.0	12875.35	175.734	227.341

†Adapted from *Federal Register*, August 9, 1973. Cost of each unit of separative work taken as typical $38.50; cost of feed per kilogram of product taken as $23.46.

product. The separative work is 4.306, with cost (4.306)($38.50) = $165.78 per kilogram. If the utility company were also to return fuel that has been used in a reactor and remains slightly enriched, a credit would be received since the amounts of natural uranium feed and separative work are smaller, as illustrated in Problem 10.6.

10.4   GAS CENTRIFUGE

This device for separating isotopes, also called the ultra-centrifuge because of the very high speeds involved, has been known since the 1940s, but only recently has become popular as a promising alternative to gaseous diffusion. It consists of a cylindrical chamber—the rotor—turning at very high speed in a vacuum (see Fig. 10.5).

The rotor is driven and supported magnetically. Gas is supplied and centrifugal force tends to compress it in the outer region, but thermal agitation tends to redistribute the gas molecules throughout the whole volume. Light molecules are favored in this effect, and their concentration is higher near the center axis. By various means, a gas flow is established that tends to carry the heavy and light isotopes to opposite ends of the rotor. Depleted and enriched streams of gas are withdrawn, as sketched in Fig. 10.6. Separation factors of 1.1 or better were obtained with centrifuges about a foot long, rotating at a rate such that the rotor

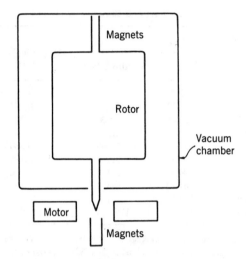

Fig. 10.5. Gas centrifuge.

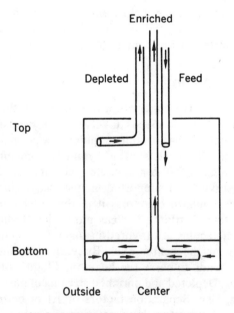

Fig. 10.6. Gas streams in centrifuge.

surface speed is 350 m/sec. The flow rate per stage of a centrifuge is much lower than that of gaseous diffusion, requiring large numbers of units in parallel. The electrical power consumption for a given capacity is lower, however, by a factor of six to ten. A gaseous diffusion plant must be very large to be efficient, and its billion dollar cost is a very great investment for one company or even a group of companies. The cost of centrifuge plants is much smaller, and they can be added as needed. The only disadvantage of the centrifuge is the lack of experience with the process. Research and development are underway in the United States, Japan, the United Kingdom, West Germany, and the Netherlands.

10.5 SEPARATION OF DEUTERIUM

The heavy isotope of hydrogen ^2_1H, deuterium, has two principal nuclear applications: (a) as low-absorption moderator for reactors, especially those using natural uranium, and (b) as a reactant in the fusion process. The differences between the chemical properties of light water and heavy water are slight, but sufficient to permit separation of ^1_1H and ^2_1H by several methods. Among these are *electrolysis*, in which the H_2O tends to be more readily dissociated, *fractional distillation*, which takes advantage of the fact that D_2O has a boiling point about 1°C higher than that of H_2O, and *catalytic exchange*, involving the passage of HD gas through H_2O to produce HDO and light hydrogen gas.

10.6 SUMMARY

The separation of isotopes requires a physical process that depends on mass. In the electromagnetic method, as used in a mass spectrograph, ions to be separated travel on circles of different radii. In the gaseous diffusion process, light molecules of a gas diffuse through a membrane more readily than do heavy molecules. The amount of enrichment in gaseous diffusion depends on the square root of ratio of the masses and is small per stage, requiring a large number of stages. By the use of material balance equations, the amount of feed can be computed, and by the use of tables of work, costs of enriching uranium for reactor fuel can be found. An alternative promising separation device is the gas centrifuge, in which gases diffuse against the centrifugal forces produced by high speeds of rotation. Several methods of separating deuterium from ordinary hydrogen are available.

10.7 PROBLEMS

10.1. (a) Show that the radi is of motion of an ion in a mass spectrograph is given by

$$r = \sqrt{\frac{2mV}{qB^2}}.$$

(b) If the masses of heavy (H) and light (L) ions are m_H and m_L, show that their separation at the plane of collection in a mass spectrograph is proportional to $\sqrt{m_H} - \sqrt{m_L}$.

10.2. The ideal separation factor for a gaseous diffusion stage is

$$r = 1 + 0.693(\sqrt{m_H/m_L} - 1).$$

Compute its value for $^{235}UF_6$ and $^{238}UF_6$, noting that $A = 19$ for fluorine.

10.3. (a) Verify that for particles of masses m_H and m_L that the number fraction f_L of the light particle is related to the weight fractions w_H and w_L by

$$f_L = \frac{n_L}{n_L + n_H} = \frac{1}{1 + \dfrac{w_H m_L}{w_L m_H}}.$$

(b) Show that the abundance ratio of numbers of particles is either

$$R = \frac{n_L}{n_H} = \frac{f_L}{1 - f_L} \quad \text{or} \quad \frac{w_L/m_L}{w_H/m_H}.$$

(c) Calculate the number fraction and abundance ratio for uranium metal that is 3% U-235 by weight. Answer: $f_L = 0.0305$, $R = 0.0315$.

10.4. Find the amount of natural uranium feed (0.711% by weight) required to produce 1 kg/day of highly enriched uranium (90% by weight), if the waste concentration is 0.25% by weight. Assume that the uranium is in the form of UF_6.

10.5. How many enriching stages are required to produce uranium that is 3% by weight, using natural UF_6 feed? Let the waste be 0.2%.

10.6. A reactor receives 3% fuel from a gaseous diffusion plant at a rate of 1 kg/day, and returns 1% fuel at 0.98 kg/day to the plant.
(a) Using Table 10.1, show that the fuel returned corresponds to a "credit" of 1.535 kg/day in feed reduction.
(b) Find the value of natural uranium feed to the gaseous diffusion plant.
(c) Find the credit in separative work for the returned uranium.
(d) Find the net separative work and the fuel enrichment cost.

11

Radiation Detectors

Measurement of radiation is required in all facets of nuclear energy—in scientific studies, in the operation of reactors for the production of power, and for protection from radiation hazard. Detectors are used to identify products of nuclear reactions, to measure neutron flux, and to determine the amount of radioisotopes in air or water. The kind of detector employed depends on several factors: the particles to be observed—electrons, gamma rays, neutrons, ions, or combinations of them; the energy of the particles; and the environment in which the detector is to be used. The type of measuring device is chosen for the intended purpose and the accuracy desired.

The demands on a detector are related to what it is we wish to know, which may be one or more of the following: (a) that there is a radiation field present, (b) the number of nuclear particles that strike the detector each second (or over some specified period of time), (c) the type of particles present and perhaps the relative number in a mixed radiation field, (d) the energy of the individual particles, (e) the instant a particle arrives at the detector. In this chapter we shall describe the important features of a few popular types of detectors. Most of them are based on the ionization produced by radiation, with the resulting currents passing through an electrical circuit and displayed on a meter or similar indicator. The number of particles of radiation arriving and noted in a given period is obtained by devices called counters.

11.1 GAS COUNTERS

Picture a gas-filled chamber with a central electrode (anode, electrically positive) and a conducting wall (cathode, negative). They are maintained at different potential, as shown in Fig. 11.1. If a charged particle or gamma ray is allowed to enter the chamber, it will produce a certain amount of ionization in the gas. The resultant positive ions and electrons are attracted toward the negative and positive surfaces, respectively. If the voltage across the tube is low, the charges merely migrate through the gas, they are collected, and a current of short duration (a pulse) passes through the resistor and the meter. More generally, amplifying circuits are required. The number of current pulses is a measure of the number of incident particles that collide in the detector, which is designated as an *ionization chamber* when operated in this mode.

If the voltage is then increased sufficiently, electrons produced by the incident radiation through ionization are able to gain enough speed to cause further ionization in the gas. Most of this action occurs near the central electrode, where the electric field is highest. The current pulses are much larger than in the ionization chamber because of the amplification effect. The current is proportional to the original number of electrons produced by the incoming radiation, and the detector is now called a *proportional counter*. One may distinguish between the passage of beta particles and alpha particles, which have widely different ability to ionize. The time for collection is very short, of the order of microseconds.

If the voltage on the tube is raised still higher, a particle or ray of any energy will set off a discharge, in which the secondary charges are so great in number that they dominate the process. The discharge stops of its own accord because of the generation near the anode of positive ions,

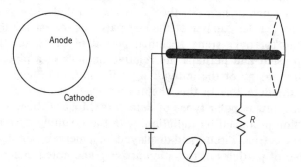

Fig. 11.1. Basic detector.

which reduce the electric field there to such an extent that electrons are not able to cause further ionization. The current pulses are then of the same size, regardless of the event that initiated them. In this mode of operation, the detector is called a *Geiger–Muller (GM) counter*. There is a short period, the "dead time," in which the detector will not count other incoming radiation. If the radiation level is very high, a correction of the observed counts to yield the "true" counts must be made, to account for the dead time. In some gases, such as argon, there is a tendency for the electric discharge to be sustained, and it is necessary to include a small amount of foreign gas or vapor, e.g., alcohol, to "quench" the discharge. The added molecules affect the production of photons and resultant ionization by them.

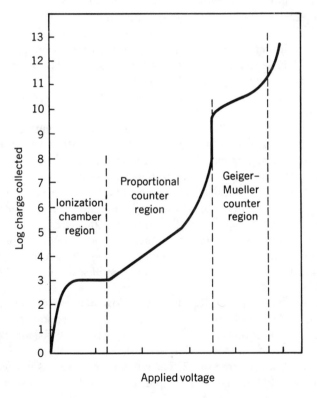

Fig. 11.2.   Collection of charge in counters. (From Raymond L. Murray, *Introduction to Nuclear Engineering*, 2nd Ed. © 1961. Reprinted by permission of Prentice-Hall, Inc., Englewood Cliffs, New Jersey.)

A qualitative distinction between the above three types of counters is displayed graphically in Fig. 11.2, which is a semilog plot of the charge collected as a function of voltage. We note that the current varies over several orders of magnitude.

11.2 NEUTRON DETECTORS

In order to detect neutrons, which do not create ionization directly, it is necessary to provide a means for generating the charges that can ionize a gas. Advantage is taken of the nuclear reaction involving neutron absorption in boron

$$_0^1n + {}_5^{10}B \rightarrow {}_2^4He + {}_3^7Li,$$

where the helium and lithium atoms are released as ions. One form of *boron counter* is filled with the gas boron trifluoride (BF_3), and operated as an ionization chamber or a proportional counter. It is especially useful for the detection of thermal neutrons since the cross section of boron-10 at 0.0253 eV is large, 3837 barns, as noted in Chapter 5. Most of the 2.8 MeV energy release goes to the kinetic energy of the product nuclei. The reaction rate of neutrons with the boron in BF_3 gas is independent of the neutron speed, as can be seen by forming the product $R = nvN\sigma_a$, where σ_a varies as $1/v$. The detector thus measures the number density n of an incident neutron beam rather than the flux. Alternatively, the metal electrodes of a counter may be coated with a layer of boron that is thin enough to allow the alpha particles to escape into the gas. The counting rate in a boron-lined chamber depends on the surface area exposed to the neutron flux. To enhance the sensitivity of detection, the counter can be constructed with a series of parallel circular plates (PCP) on a rod, while alternating plates are connected to the case, as shown in Fig. 11.3. A similar arrangement is found in gamma ray sensitive detectors, where the gammas produce secondary electrons in a metal wall, preferably of large Z value.

The *fission chamber* is often used for slow neutron detection. A thin layer of U-235, with high thermal neutron cross section, 681 barns, is deposited on the cathode of the chamber. Energetic fission fragments produced by neutron absorption traverse the detector and give the necessary ionization. Uranium-238 is avoided because it is not fissile with slow neutrons and because of its stopping effect on fragments from U-235 fission.

Neutrons in the thermal range can be detected by the radioactivity induced in a substance in the form of small foil or thin wire. Examples are

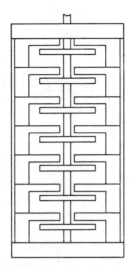

Fig. 11.3.   Parallel circular plate ionization chamber.

manganese $^{55}_{25}$Mn, with a 13-barn cross section at 2200 m/sec, which becomes $^{56}_{25}$Mn with half-life 2.58 h; and dysprosium $^{164}_{66}$Dy, 1000 barns, becoming $^{165}_{66}$Dy, half-life 140 min. For detection of neutrons slightly above thermal energy, materials with a high resonance cross section are used, e.g., indium, with a peak at 1.45 eV. To separate the effects of thermal neutron capture and resonance capture, comparisons are made between measurements made with thin foils of indium and those of indium covered with cadmium. The latter screens out low-energy neutrons (below 0.5 eV) and passes those of higher energy.

For the detection of fast neutrons, up in the MeV range, the *proton recoil* method is used. We recall from Chapter 5 that the scattering of a neutron by hydrogen results in an energy loss, which is an energy gain for the proton. Thus a hydrogenous material such as methane (CH_4) or H_2 itself may serve as the counter gas. The energetic protons play the same role as did alpha particles and fission fragments in the counters discussed previously. Nuclear reactions such as $^3_2$He (n, p) $^3_1$H can also be employed to obtain detectable charged particles.

11.3   SCINTILLATION COUNTERS

The name of this detector comes from the fact that the interaction of a particle with some materials gives rise to a scintillation or flash of light.

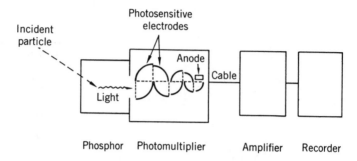

Fig. 11.4.   Scintillation detection system.

The basic phenomenon is familiar—many substances can be stimulated to glow visibly on exposure to ultraviolet light, and the images on a color television screen are the result of electron bombardment. Molecules of materials classed as phosphors are excited to states that release their energy statistically, in close analogy to the radioactive decay process. The substances used in the scintillation detector are inorganic, e.g., sodium iodide or lithium iodide, or organic, in one of various forms—crystalline, plastic, liquid, or gas.

The amount of light released when a particle strikes a phosphor is proportional to the energy deposited, and thus makes the detector especially useful for the determination of particle energies. Since charged particles have a short range, most of their energy appears in the substance. Gamma rays also give rise to an energy deposition through electron recoil in both the photoelectric effect and Compton scattering, and through the pair production-annihilation process. A schematic diagram of a detector system is shown in Fig. 11.4. Some of the light released in the phosphor is collected in the photomultiplier tube, which consists of a set of electrodes with photosensitive surfaces. When a photon strikes the surface, an electron is emitted by the photoelectric effect, it is accelerated to the next surface where it dislodges more electrons, and so on, and a multiplication of current is achieved. An amplifier then increases the electrical signal to a level convenient for counting or recording.

11.4   SOLID-STATE DETECTORS

The use of a solid rather than a gas in a detector has the advantage of compactness, since the range of nuclear particles is very short in solids. In addition, when the detector is composed of semiconductor materials, great accuracy in the measurement of energy and time of arrival is achieved. There are many similarities between gas and solid detectors, but the mechanism of ion motion is quite different. We can visualize a crystal semiconductor, such as silicon or germanium, as a regular array of fixed atoms, with some freedom of electron motion in the lattice. An incident charged particle can detach an electron from an atom and cause it to leave the vicinity. The removal of the electron causes a vacancy or "hole," which acts effectively as a positive charge. The electron-hole *pair* produced is analogous to negative and positive ions in a gas. Electrons can migrate through the material or be carried along by an electric field, while the holes "move" as electrons are successively exchanged with neighboring atoms.

The electrical conductivity of a semiconductor is very sensitive to the presence of certain impurities. Consider silicon, which has a chemical valence of 4, i.e., 4 electrons in the outer shell. The introduction of small amounts of an element with valence 5, such as phosphorous or arsenic, provides additional negative charge, and the resulting material is classed as *n*-type silicon. If, instead, an element with valence 3 is added, such as boron or gallium, there is a scarcity of electrons or an excess of positive holes, and the material is called *p*-type silicon. When two layers of *n*-type and *p*-type materials are put in close contact and a voltage is applied to the outside surfaces as in Fig. 11.5, electrons are drawn one way, holes the other, leaving a neutral or "depleted" region between. Most of the voltage drop occurs across the neutral zone, since it is nearly a perfect insulator. The depleted region is very sensitive to radiation. The electron-hole pairs resulting from an incident particle are swept out by the

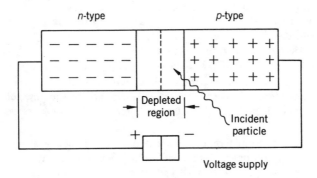

Fig. 11.5.   Solid state *n–p* junction detector.

high internal electric field and register as a current pulse. The ability of an
$n-p$ junction detector to measure the energy of nuclear particles
accurately is the result of the fact that the energy required to create a pair
in silicon or germanium is about 3 eV (in comparison with about 30 eV to
create an ion pair in a gas). A photon of say 0.1 MeV energy gives rise to a
very large number of charge pairs, and thus statistical accuracy is higher.
In addition, the time required for charges to be collected is extremely
short, about one billionth of a second, permitting precise measurement of
the time of the counting event.

11.5 SUMMARY

The detection of radiation and the measurement of its properties is
required in all aspects of the nuclear field. In gas counters, the ionization
produced by incoming radiation is collected. Dependent on the voltage
applied between electrodes, counters detect all particles or distinguish
between types. Neutrons are detected indirectly by the products of
nuclear reactions—for slow neutrons by absorption in boron or uranium,
for fast neutrons by scattering in hydrogen. Scintillation counters release
measurable light on bombardment by charged particles or gamma rays,
while solid-state detectors take advantage of the sensitivity of semicon-
ductors to a disturbance of the charge balance.

11.6 PROBLEMS

11.1. (a) Find the number density of molecules of BF_3 in a detector of 1-in.
diameter to be sure that 90% of the thermal neutrons incident along a diameter are
caught (σ_a for natural boron is 759 barns).
(b) How does this compare with the number density for the gas at atmospheric
pressure, with density 3.0×10^{-3} g/cm^3?
(c) Suggest ways to achieve the high efficiency desired.

11.2. An incident particle ionizes helium to produce two electrons and an alpha
particle halfway between two parallel plates with potential difference between
them. If the gas pressure is very low, estimate the ratio of the times elapsed until
the charges are collected, t_e/t_α. Discuss the effect of collisions on the collection
time.

11.3. We collect a sample of gas suspected of containing a small amount of
radioiodine, half-life 8 days. If we observe in a period of 1 day a total count of
50,000 in a counter that detects all radiation emitted, how many atoms were
initially present?

11.4. In a gas counter, the potential difference at any point r between a central wire of radius r_1 and a concentric wall of radius r_2 is given by

$$V = V_0 \frac{\ln (r/r_1)}{\ln (r_2/r_1)},$$

where V_0 is the voltage across the tube. If $r_1 = 1$ mm and $r_2 = 1$ cm, what fraction of the potential difference exists within a millimeter of the wire?

11.5. How many electrodes would be required in a photomultiplier tube to achieve a multiplication of one million if one electron releases four electrons?

12

Neutron Chain Reactions

The possibility of a chain reaction involving neutrons and a nuclear fuel such as uranium is dependent on the number of neutrons produced per absorption, η, as discussed in Chapter 7. Its value must be more than one because of inevitable losses of neutrons. However, to achieve a self-sustaining chain reaction, one in which no neutrons need to be supplied, a certain amount of uranium must be brought together, the "critical" mass. In order to appreciate this requirement, we can visualize an experiment in which we can assemble various amounts of U-235. In effect, we will be building a nuclear reactor but will not consider all of the materials and components. We ignore parts that are present for structural support, to provide control, to permit the extraction of energy, or to give protection from radiation. The minimum ingredients are a nuclear fuel and at least one neutron to start the process.

12.1 THE SELF-SUSTAINING CHAIN REACTION

If we have only one atom of U-235, and it is bombarded by a neutron to induce fission, the resultant neutrons are able to do nothing further, there being no more fuel. Instead, we form a small sphere of uranium of volume, say, $1\,cm^3$ containing about $19\,g$. The number of nuclei is adequate for a very long sequence of fission events, but on introducing a neutron, the series of reactions soon ends because of loss from the surface of the sphere. Only if we were to supply neutrons continually to make up for this "leakage" could we keep the reaction going. Such an assembly is called "subcritical." However, if we bring together about $50\,kg$ of U-235 metal in spherical form, the rate of neutron production is

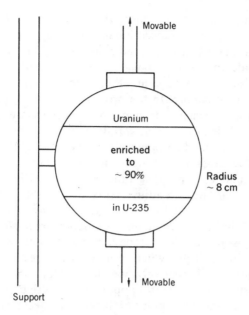

Fig. 12.1.   Fast metal assembly "Godiva."

sufficient to balance leakage losses, and an outside supply of neutrons is not needed. In terms of neutrons, the assembly is self-sustaining or "critical." The amount of nuclear fuel is called the critical mass, its volume is the critical size. Figure 12.1 shows the uranium metal assembly Godiva, so named because it was "bare," i.e., had no surrounding materials. This reactor, composed of highly enriched uranium, was used for many years for test purposes at Los Alamos, New Mexico.

If we add still more uranium beyond the 50 kg required for criticality, more neutrons are produced than are used, the neutron population grows, and the reactor is "supercritical." Early nuclear weapons consisted of two hemispheres of uranium, each subcritical and unable separately to sustain a chain reaction. When suddenly brought together they formed a highly supercritical mass, in which the rapid growth in the numbers of neutrons, fission events, and energy developed produced a violent explosion.

12.2 MULTIPLICATION FACTORS

We may describe any arrangement of fuel material by a single number, the effective multiplication factor k (or k_{eff}), as being the *net* number of neutrons produced (accounting for all possible losses) per initial neutron.

If k is less than 1, the system is subcritical; if k is equal to 1, it is critical; and if k is greater than 1, supercritical. The design and operation of all reactors is focused on k or on related quantities.† The choice of materials and size is made to assure a balance between neutron production by fission and losses by capture in other elements or leakage from the boundaries of the assembly. In the process of bringing parts together in what is called a critical experiment, observations are made that give estimates of k. During operation, variations in k are made as needed by adjustments of neutron-absorbing rods or dispersed chemicals that cause increases or decreases in the neutron population. Eventually, in the operation of a reactor for a long time, enough fuel is consumed that k goes below 1 regardless of adjustments of control materials, and the reactor must be shut down for refueling.

Two views of the growth of human population are analogous to neutron multiplication. A person born today in the United States has a life expectancy of about 70 yr, which is a statistical result of past data on individual longevity. Alternatively, we may say that the birthrate exceeds the death rate such that there is a population growth of 2% per year. The first view involves the probability of survival or death of the individual; the second view compares rates of competing processes that affect the total population.

Similarly, we can focus attention on a typical neutron that starts its life in fission, and has various chances of dropping out of the cycle because of leakage and absorption in other materials besides fuel. Or, at a given time we can form the sum of the reaction rates for the processes of neutron absorption, fission with neutron yield, and leakage in order to find out if the neutron population is increasing, is steady, or is decreasing. Each method has its merits for purposes of discussion or analysis.

The statistical approach involves the observation of many histories and deducing averages. Let us look at the possible behaviors of several fission neutrons, using the uranium metal reactor for reference. As in Fig. 12.2(a), a neutron may escape on first flight from the sphere, since mean free paths are rather long. Another (b) may make a scattering collision and then escape. Others may collide and be absorbed either (c) to form U-236 or (d) to give rise to fission, the latter case yielding three neutrons in the case shown. Several collisions may occur before leakage or absorption takes place. A "flow diagram" as in Fig. 12.3 is useful to describe the fates. The boxes represent processes, the circles the numbers of neutrons.

†Such as $\delta k = k - 1$ or $\delta k / k$ called reactivity, symbolized by ρ.

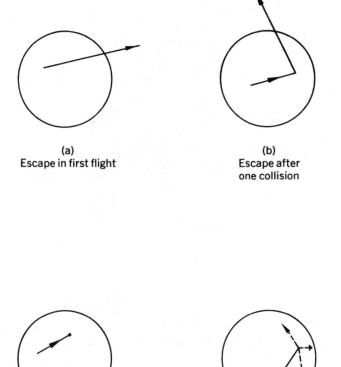

(a)
Escape in first flight

(b)
Escape after
one collision

(c)
Absorption in U-235
to form U-236

(d)
Absorption with
neutron production

Fig. 12.2. Neutron histories.

The fractions of absorbed neutrons that form U-236 and that cause fission, respectively, are the ratios of the cross section for capture σ_c and fission σ_f to that for absorption σ_a. The average number of neutrons produced by fission is ν, where for simplicity we omit the bar signifying average. Now let η be the combination $\nu\sigma_f/\sigma_a$, and note that it is the number of neutrons per absorption in uranium. Thus letting $\mathscr{L}$ be the

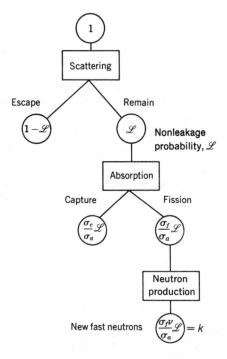

Fig. 12.3.   Neutron cycle in metal assembly.

fraction *not* escaping by leakage,

$$k = \eta \mathscr{L}.$$

The system is critical if $k = 1$, or $\eta \mathscr{L} = 1$. Measurements show that η is around 2.3 for fast neutrons, thus $\mathscr{L}$ must be $1/2.3 = 0.43$, which says that as many as 43% of the neutrons must remain in the sphere, while no more than 57% escape through its boundary.

12.3   NEUTRON BALANCES

Let us now look at the same uranium metal system from the other viewpoint, that of reaction rates. A critical reactor is one in which there is a balance of processes:

$$\text{production} = \text{absorption} + \text{leakage}.$$

This word equation relates the rates of neutron generation by fission, of removal from the system by absorption, and loss through the boundaries. The statement of balance can be put in a form that expresses the relationship of materials content and the size and shape of the reactor, according to the following logic. Suppose that the reactor is pictured as a single region of volume V with no surrounding material. The neutron flux may be expected to be large in the center and small near the edges because of the general outward flow of neutrons. Let the average flux over the core be $\bar{\phi}$ and the macroscopic cross section for absorption be Σ_a. Then the rate of *absorption* is $\bar{\phi}\Sigma_a V$. The production term can be formed by analogy. If the fission cross section is Σ_f, the fission rate over the core is $\bar{\phi}\Sigma_f V$, and if each fission yields ν neutrons, the rate of *production* is $\bar{\phi}\Sigma_f V\nu$. The rate of *leakage* is found to be $\bar{\phi}DB^2V$, where D is the diffusion coefficient defined in Chapter 5 and where a new quantity B^2, called *buckling*, is dependent on the geometric features of the reactor. For a sphere of radius R, the buckling is $B^2 = (\pi/R)^2$, with other formulas for various shapes.† If we use these in our balance equation and simplify, we find

$$\Sigma_f\nu = \Sigma_a + DB^2.$$

This formula represents a *critical condition*, relating material content and geometric features that must be compatible if the system is to be self-sustaining.

   Let us calculate the critical size of a sphere of pure U-235. For metal of density 19 g/cm^3, $N_u = 0.048 \text{ cm}^{-3}$ (in units of 10^{24}), and with cross sections $\sigma_f = 1.40$ barns, $\sigma_a = 1.65$ barns, $\sigma_t = 6.8$ barns, and $\nu = 2.6$, we find $\Sigma_f = 0.0672 \text{ cm}^{-1}$, $\Sigma_a = 0.0792 \text{ cm}^{-1}$, $\Sigma_t = 0.326 \text{ cm}^{-1}$, $\lambda_t = 1/\Sigma_t = 3.07 \text{ cm}$, $D = \lambda_t/3 = 1.02 \text{ cm}$, $B^2 = (\Sigma_f\nu - \Sigma_a)/D = 0.0936 \text{ cm}^{-2}$, and $R = \pi/B = 10.3 \text{ cm}$. A slight correction is required to take proper account of the shape of the flux near the boundary. The calculated radius is smaller by an amount $0.71 \lambda_t$, or 2.2 cm for this case. We thus predict the sphere radius to be 8.1 cm, which is in reasonable agreement with the actual size of the Godiva reactor. The core volume is $V = \frac{4}{3}\pi R^3 = 2.2 \times 10^3 \text{ cm}^3$, and using the given density, the calculated critical mass is 42 kg.

   The critical condition $\Sigma_f\nu = \Sigma_a + DB^2$ enables us to make a variety of reactor calculations. If the size were known, the necessary properties of fuel can be found, the reverse of the case just examined. As another example, suppose we substitute a fuel that releases more neutrons in fission. To keep the reactor critical, the value of $B^2 = (\pi/R)^2$ must be

†For a rectangular parallelepiped of sides L, W, and H, it is $B^2 = (\pi/L)^2 + (\pi/W)^2 + (\pi/H)^2$; thus for a cube of side S, it is $B^2 = 3(\pi/S)^2$; for a circular cylinder of height H and radius R, $B^2 = (2.405/R)^2 + (\pi/H)^2$.

raised, by decreasing the radius R. If we add a strong absorber such as boron, Σ_a will increase, B^2 must be smaller, and thus R must increase to compensate. The reader can investigate the effect of uniform expansion which reduces the number density of all materials and also causes the radius to change (see Problem 12.3). It is easy to show that our critical condition is identical to $1 = \eta\mathcal{L}$ if $\eta = (\Sigma_f\nu/\Sigma_a)_u$ and $\mathcal{L} = 1/(1 + DB^2/\Sigma_a)$.

12.4 REACTOR POWER

The power developed by a reactor is a quantity of great interest for practical reasons. Power is related to the neutron population, and also to the mass of fissile material present. First, let us look at a typical cubic centimeter of the reactor, containing N fuel nuclei, each with cross section for fission σ_f at the typical neutron energy of the reactor, corresponding to neutron speed v. Suppose that there are n neutrons in the volume. The rate at which the fission reaction occurs is thus $R_f = nvN\sigma_f$ fissions per second. If each fission produces an energy w, then the power per unit volume is $p = wR_f$. For the whole reactor, of volume V, the rate of production of thermal energy is $P = pV$. If we have an average flux $\bar{\phi} = nv$ and a total number of fuel atoms $N_T = NV$, the total reactor power is seen to be

$$P = \bar{\phi}N_T\sigma_f w.$$

Thus we see that the power is dependent on the product of the number of neutrons and the number of fuel atoms. A high flux is required if the reactor contains a small amount of fuel, and conversely. All other things equal, a reactor with a high fission cross section can produce a required power with less fuel than one with small σ_f. We recall that σ_f decreases with increasing neutron energy. Thus for given power P, a "fast" reactor, one operating with neutron energies principally in the vicinity of 1 MeV, requires either a much larger flux or a larger fissile fuel mass than does the "thermal" reactor, with neutrons of energy around 0.1 eV.

The power developed by most familiar devices is closely related to fuel consumption. For example, a large car generally has a higher gasoline consumption rate than a small car, and more gasoline is used in operation at high speed than at low speed. In a reactor, it is necessary to add fuel very infrequently because of the very large energy yield per pound, and the fuel content remains essentially constant. From the formula relating power, flux, and fuel, we see that the power can be readily raised or lowered by changing the flux. By manipulation of control rods, the neutron population is allowed to increase or decrease to the proper level.

Power reactors used to generate electricity produce about 3000 megawatts of thermal power (MWt), and with an efficiency of conversion of around $\frac{1}{3}$, give 1000 MW of electrical power (1000 MWe).

12.5 MULTIPLICATION IN A THERMAL REACTOR

The presence of large amounts of neutron-moderating material such as water in a reactor greatly changes the neutron distribution in energy. Fast neutrons slow down by means of collisions with light nuclei, with the result that most of the fissions are produced by low-energy (thermal) neutrons. Such a system is called a "thermal" reactor in contrast with a system without moderator, a "fast" reactor, operating principally with fast neutrons. The cross sections for the two energy ranges are widely different, as noted in Problem 12.6. Also, the neutrons are subject to being removed from the multiplication cycle during the slowing process by strong resonance absorption in elements such as U-238. Finally, there is a competition for the neutrons between fuel, coolant, structural materials, fission products, and control absorbers.

The description of the multiplication cycle is somewhat more complicated than that for a fast metal assembly, as seen in Fig. 12.4. The set of reactor parameters are (a) the fast fission factor ϵ, representing the immediate multiplication because of fission at high neutron energy, mainly in U-238; (b) the fast nonleakage probability $\mathscr{L}_f$, being the fraction remaining in the core during neutron slowing; (c) the resonance escape probability p, the fraction of neutrons *not* captured during slowing; (d) the thermal nonleakage probability $\mathscr{L}_t$, the fraction of neutrons remaining in the core during diffusion at thermal energy; (e) the thermal utilization f, the fraction of thermal neutrons absorbed in fuel; and (f) the reproduction factor η, as the number of new fission neutrons per absorption in fuel. At the end of the cycle starting with one fission neutron, the number of fast neutrons produced is seen to be $\epsilon p f \eta \mathscr{L}_f \mathscr{L}_t$, which may be also labeled k, the effective multiplication factor. It is convenient to group four of the factors to form $k_\infty = \epsilon p f \eta$, the so-called "infinite multiplication factor" which would be identical to k if the medium were infinite in extent, without leakage. If we form a composite nonleakage probability $\mathscr{L} = \mathscr{L}_f \mathscr{L}_t$, then we may write

$$k = k_\infty \mathscr{L}.$$

For a reactor to be critical, k must equal 1, as before.

To provide some appreciation of the sizes of various factors, let us calculate the values of the composite quantities for a typical thermal

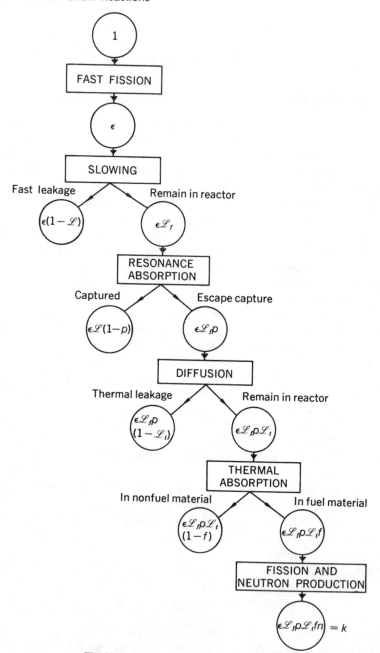

Fig. 12.4. Neutron cycle in thermal reactor.

power reactor, for which $\epsilon = 1.03$, $p = 0.71$, $\mathscr{L}_f = 0.97$, $\mathscr{L}_t = 0.99$, $f = 0.79$, and $\eta = 1.8$. Now $k_\infty = (1.03)(0.71)(1.8)(0.79) = 1.04$, $\mathscr{L} = (0.97)(0.99) = 0.96$, and $k = (1.04)(0.96) = 1.00$. For this example, the various parameters yield a critical system. In the following chapter we shall describe the physical construction of typical thermal reactors.

12.6 SUMMARY

A self-sustaining chain reaction involving neutrons and fission is possible if the accumulation of nuclear fuel is large enough, i.e., a critical mass is brought together. The value of the multiplication factor k indicates whether a reactor is subcritical (< 1), critical ($= 1$) or supercritical (> 1). There is an analogy between neutron and human populations; and in both, the history of a typical individual or the rate of change of numbers can be studied. The critical condition formula permits the calculation of reactor size for given materials content (or vice versa), and the study of effects of changes. The reactor power, dependent on the product of flux and fuel atoms, is readily adjustable. A thermal reactor contains moderator and operates on slowed neutrons.

12.7 PROBLEMS

12.1. Calculate the reproduction factor η for fast neutrons in pure U-235, using values of σ_f, σ_a, and ν cited in the text.

12.2. Calculate $\mathscr{L}$ and show that $\eta\mathscr{L} = 1$ is met numerically by the calculated size of the U-235 metal assembly.

12.3. Using the critical condition for a bare reactor, written as

$$\Sigma_f \nu - \Sigma_a = DB^2,$$

where for a cube $B^2 = 3(\pi/S)^2$, examine the effect on criticality of uniform thermal expansion. The cross sections and diffusion coefficient are dependent on fuel number density N_u, which varies inversely with volume $V = S^3$.

12.4. If the power developed by the Godiva-type reactor is 100 W, and the total mass of fuel is 50 kg, what average flux is implied? Note that the energy per fission is $w \simeq 3.2 \times 10^{-11}$ W-sec.

12.5. Find the multiplication factors k_∞ and k for a thermal reactor for which $\epsilon = 1.05$, $p = 0.80$, $\mathscr{L}_f = 0.75$, $\mathscr{L}_t = 0.95$, $f = 0.85$, and $\eta = 2.08$. Evaluate the reactivity ρ.

12.6. The number of neutrons per absorption in partly enriched uranium depends on the properties of the two isotopes (1) U-235 and (2) U-238 as follows:

$$\eta = \frac{N_1\sigma_{f1}\nu_1 + N_2\sigma_{f2}\nu_2}{N_1\sigma_{a1} + N_2\sigma_{a2}}.$$

Compute the value of η for uranium of 3% U-235 by weight, $N_1/N_2 = 0.0315$, (Problem 10.3) for two cases: (a) a fast reactor, and (b) a thermal reactor, using the table of data below. Comment on the results.

	Thermal	Fast
σ_{f1}	582	1.40
σ_{a1}	681	1.65
σ_{f2}	0	0.095
σ_{a2}	2.70	0.255
ν_1	2.43	2.6
ν_2	0	2.6

13

Nuclear Reactor Concepts

Although the only requirement for a neutron chain reaction is a sufficient amount of a fissionable element, many combinations of materials and arrangements can be used to construct an operable nuclear reactor. Several different types or concepts have been devised and tested over the period since 1942, when the first reactor started operation, just as various kinds of engines have been used—steam, internal combustion, reciprocating, rotary, jet, etc. Experience with the individual reactor concepts has led to the selection of a few that are most suitable, using criteria such as economy, reliability, and ability to meet performance demands.

In this chapter we shall identify these important reactor features, compare several concepts, and then focus attention on the components of one specific power reactor type. We shall then examine the processes of fuel consumption and control in a power reactor.

13.1 REACTOR CLASSIFICATION

A general classification scheme for reactors has evolved that is related to the distinguishing features of the reactor types. These features are listed below.

(a) Purpose

The majority of reactors in operation or under construction have as purpose the generation of large blocks of commercial electric power. Others serve training or radiation research needs, and many provide propulsion power for submarines. Available also are tested reactors for commercial surface ships and for spacecraft. At various stages of development of a new concept, such as the breeder reactor, there will be constructed a prototype reactor, one which tests feasibility, and a demonstration reactor, one that evaluates commercial possibilities.

(b) Neutron Energy

A fast reactor is one in which most of the neutrons are in the energy range 0.1–1 MeV, below but near the energy of neutrons released in fission. The neutrons remain at high energy because there is relatively little material present to cause them to slow down. In contrast, the thermal reactor contains a good neutron moderating material, and the bulk of the neutrons have energy in the vicinity of 0.1 eV.

(c) Moderator and Coolant

In some reactors, one substance serves two functions—to assist in neutron slowing and to remove the fission heat. Others involve one material for moderator and another for coolant. The most frequently used materials are listed below:

Moderators	Coolants
light water	light water
heavy water	carbon dioxide
graphite	helium
beryllium	liquid sodium

The condition of the coolant serves as a further identification. The *pressurized water reactor* provides high-temperature water to a heat exchanger that generates steam, while the *boiling water reactor* supplies steam directly.

(d) Fuel

Uranium with U-235 content varying from natural uranium ($\approx 0.7\%$) to slightly enriched ($\approx 3\%$) to highly enriched ($\approx 90\%$) is employed in various reactors, with the enrichment depending upon what other absorbing materials are present. The fissile isotopes $^{239}_{94}\text{Pu}$ and $^{233}_{92}\text{U}$ are produced and consumed in reactors containing significant amounts of U-238 or Th-232. Plutonium serves as fuel for fast breeder reactors and can be recycled as fuel for thermal reactors. The fuel may have various physical forms—a metal, or an alloy with a metal such as aluminum, or a compound such as the oxide UO_2 or carbide UC.

(e) Arrangement

In most modern reactors, the fuel is isolated from the coolant in what is called a *heterogeneous* arrangement. The alternative is a homogeneous mixture of fuel and moderator or fuel and moderator-coolant.

(f) Structural Materials

The functions of support, retention of fission products, and heat conduction are provided by various metals. The main examples are aluminum, stainless steel, and zircaloy, an alloy of zirconium and zinc.

By placing emphasis on one or more of the above features of reactors, reactor concepts are identified. Some of the more widely used or promising power reactor types are the following:

PWR (pressurized water reactor), a thermal reactor with light water at high pressure (2200 psi) and temperature (600°F) serving as moderator-coolant, and a heterogeneous arrangement of slightly enriched uranium fuel.

BWR (boiling water reactor), similar to the PWR except that the pressure and temperature are lower (1000 psi and 550°F).

HTGR (high temperature gas-cooled reactor), using graphite moderator, highly enriched uranium with thorium, and helium coolant (1430°F and 600 psi).

CANDU (Canadian deuterium uranium) using heavy water moderator

Table 13.1. Power Reactor Materials.

	Pressurized water (PWR)	Boiling water (BWR)	Natural uranium heavy water (CANDU)	High temp. gas-cooled (HTGR)	Liquid metal fast breeder (LMFBR)
Fuel form	UO_2	UO_2	UO_2	UC_2, ThC_2	PuO_2, UO_2
Enrichment	3% U-235	2.5% U-235	0.7% U-235	93% U-235	15 wt.% Pu-239
Moderator	water	water	heavy water	graphite	none
Coolant	water	water	heavy water	helium gas	liquid sodium
Cladding	zircaloy	zircaloy	zircaloy	graphite	stainless steel
Control	B_4C or Ag–In–Cd rods	B_4C crosses	moderator level	B_4C rods	tantalum or B_4C rods
Vessel	steel	steel	steel	prestressed concrete	steel

and natural uranium fuel that can be loaded and discharged during operation.

LMFBR (liquid metal fast breeder reactor), with no moderator, liquid sodium coolant, and plutonium fuel, surrounded by natural or depleted uranium.

Table 13.1 amplifies on the principal features of the five main power reactor concepts.

13.2 POWER REACTORS

The large-scale reactors used for the production of thermal energy that is converted to electrical energy are much more complex than the fast assembly described in Chapter 12. To illustrate, we can identify the components and their functions in a modern pressurized water reactor. Figure 13.1 gives some indication of the sizes of the various parts.

The fuel in the PWR consists of cylindrical pellets of slightly enriched (3% U-235) uranium oxide (UO_2) of diameter around $\frac{3}{8}$ in. and length $\frac{3}{4}$ in. A zircaloy tube about 12 ft long, 0.025 in. wall thickness, is filled with the pellets and sealed to form a fuel rod (or pin). The metal container serves to provide support for the column of pellets, to provide cladding that retains radioactive fission products, and to protect the fuel from interaction with the coolant. About 200 of the fuel pins are grouped in a bundle called a fuel element of about 8 in. on a side, and about 180 elements are assembled in an approximately cylindrical array to form the reactor *core*. This structure is mounted on supports in a steel *pressure vessel* of outside diameter about 16 ft, height 40 ft, and walls up to 12 in. thick. *Control rods*, consisting of an alloy of cadmium, silver, and indium, provide the ability to change the amount of neutron absorption. The rods are inserted in some vacant fuel pin spaces and magnetically connected to drive mechanisms. On interruption of magnet current, the rods enter the core through the force of gravity. The pressure vessel is filled with light water, which serves as neutron moderator, as coolant to remove fission heat, and as *reflector*, the layer of material surrounding the core that helps prevent neutron escape. The water also contains in solution the compound boric acid H_3BO_3, which strongly absorbs neutrons in proportion to the number of boron atoms and thus inhibits neutron multiplication, i.e., "poisons" the reactor. The term *soluble poison* is often used to identify this material, the concentration of which can be adjusted during reactor operation. To keep the reactor critical as fuel is consumed, the boron content is gradually reduced. A *shield* of concrete

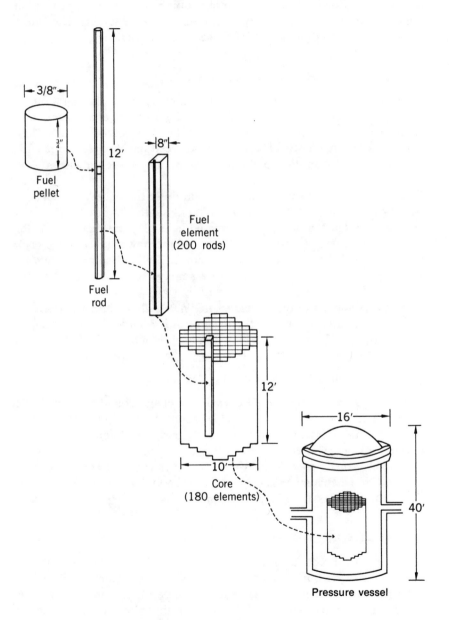

Fig. 13.1.   Reactor construction.

surrounds the pressure vessel and other equipment to provide protection against neutrons and gamma rays from the nuclear reactions. The shield also serves as an additional barrier to the release of radioactive materials.

13.3 CONSUMPTION OF NUCLEAR FUELS

The generation of energy from nuclear fuels is unique in that a rather large amount of fuel must be present at all times for the chain reaction to continue. (In contrast, an automobile will operate even though its gasoline tank is practically empty.) There is a subtle relationship between reactor fuel and other quantities such as consumption, power, neutron flux, criticality, and control.

The first and most important consideration is the energy production, which is directly related to fuel consumption. Let us simplify the situation by assuming that the only fuel consumed is U-235, and that the reactor operates continuously and steadily at a definite power level. Since each atom "burned," i.e., converted into either U-236 or fission products by neutron absorption, has an accompanying energy release, we can find the amount of fuel that must be consumed in a given period.

Let us examine the fuel usage in a PWR with initial enrichment in U-235 of 3%. Suppose that the thermal power is 3000 MW and the reactor operates for 1 yr. Using the convenient rule of thumb that 1.3 g of U-235 is burned for each megawatt-day of energy, the weight of fuel used is

$$(3000 \text{ MW})(365 \text{ days})(1.3 \text{ g/MW-day}) = 1.4 \times 10^6 \text{ g}.$$

Now each gram of U-235 at 3% enrichment costs around $10 (see Table 10.1). The cost of the fuel consumed is around 14 million dollars. Adding the cost of fuel fabrication and inventory charges will double this figure. A typical efficiency of conversion of thermal energy to electrical energy is $\frac{1}{3}$, so the electrical power is 1000 MW. Over the year (8760 hr) the energy delivered to the customers is 8.76×10^9 kWh, and thus the fuel cost is, per kilowatt-hour, $0.0032, 0.32¢, or 3.2 mills.

13.4 REACTOR CONTROL

Since no fuel is added during the operating cycle of the order of a year, the amount to be burned must be installed at the beginning. First, the amount of uranium needed to achieve criticality is loaded into the reactor. If then the "excess" is added, it is clear that the reactor would be supercritical unless some compensating action were taken. In the PWR, the excess fuel reactivity is reduced by the inclusion of control rods and boron solution.

The reactor is brought to full power and operating temperature and pressure by means of rod position adjustments. Then, as the reactor operates and fuel begins to burn out, the concentration of boron is reduced. By the end of the cycle, the extra fuel is gone, all of the available control absorption has been removed, and the reactor is shut down for refueling. The trends in fuel and boron are shown in Fig. 13.2, neglecting the effects of certain fission product absorption and plutonium production. The graph represents a case in which the power is kept constant. The fuel content thus linearly decreases with time. Such operation characterizes a reactor that provides a "base load" in an electrical generating system that also includes fossil fuel plants and hydroelectric stations.

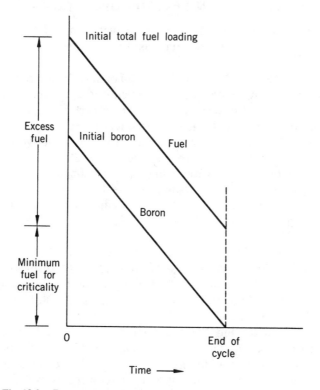

Fig. 13.2. Reactor control during fuel consumption in power reactor.

The power level in a reactor was shown in Chapter 12 to be proportional to neutron flux. However, in a reactor that experiences fuel consumption the flux must increase in time, since the power is proportional also to the fuel content.

The amount of control absorber required at the beginning of the cycle is proportional to the amount of excess fuel added to permit burnup for power production. For example, if the fuel is expected to go from 3% to 1.5% U-235, a boron atom number density in the moderator coolant is about 1.0×10^{-4} (in units of 10^{24}). For comparison, the number of water molecules per cubic centimeter is 0.0334. The boron content is usually expressed in parts per million (i.e., micrograms of an additive per gram of diluent). For our example, using 10.8 and 18.0 as the molecular weights of boron and water, there are $10^6(10^{-9})(10.8)/(0.0334)(18.0) = 1800$ ppm.

13.5   OTHER EFFECTS ON REACTOR OPERATION

The description of the reactor process just completed is somewhat idealized. Several other phenomena must be accounted for in design and operation.

If a reactor is fueled with natural uranium or slightly enriched uranium, the generation of plutonium tends to extend the cycle time. The fissile Pu-239 helps maintain criticality and provides part of the power.

Neutron absorption in the fission products has an effect on control requirements. The most important of these is a radioactive isotope of xenon, Xe-135, which has a cross section at 0.0253 eV of around 2.6 *million* barns. Its yield in fission is high, $y = 0.06$, meaning that for each fission, one obtains 6% as many atoms of Xe-135. In steady operation at high neutron flux, its rate of production is equal to its consumption by neutron absorption. Hence

$$N_X \sigma_{aX} = N_F \sigma_{fF} y.$$

Using the ratio σ_f / σ_a for U-235 of 0.85, we see that the absorption rate of Xe-135 is around $(0.85)(0.06) = 0.05$ times that of the fuel itself. This factor is about 0.04 if the radioactive decay ($t_H = 9.13$ hr) of xenon-135 is included (see Problem 13.3).

It might appear from Fig. 13.2 that the reactor cycle could be increased to as long a time as desired merely by adding more U-235 at the beginning. There are limits to such additions, however. First, the more the excess fuel that is added, the greater must be the control by rods or soluble poison. Second, radiation and thermal effects on fuel and cladding materials increase with life. The amount of allowable total energy extracted from the uranium, including all fissionable isotopes, is expressed as the number of megawatt-days per metric ton (MWd/tonne).† We can calculate its value for the year's operation by noting that the initial U-235 loading was 2800 kg, twice that burned; with an enrichment of 0.03, the *uranium* content was 2800/0.03 = 93,000 kg or 93 tonnes. Using the energy yield of (3000 MW)(365 days) ≅ 1,100,000 MWd, we find 12,000 MWd/tonne. It is desirable to seek larger values of this quantity, in order to prolong the cycle and thus minimize the cost of fuel reprocessing and fabrication.

13.6 SUMMARY

Reactors are classified according to their important features such as purpose, neutron energy, moderator and coolant, fuel, arrangement, and structural material. The principal types are the pressurized water reactor, the boiling water reactor, the high-temperature gas-cooled reactor, and the liquid metal fast breeder reactor. Excess fuel is added to a reactor initially to take care of burning during the operating cycle, with adjustable control absorbers present to maintain criticality. Account must be taken of fission product absorbers, especially the high cross section Xe-135, and of limitations on the energy release set by thermal and radiation effects.

13.7 PROBLEMS

13.1. How many individual fuel pellets are there in the PWR reactor described in the text? Assuming a density of uranium oxide of 10 g/cm³, estimate the total mass of uranium and U-235 in the core in kilograms. What is the initial fuel cost? (see Chapter 10).

13.2. How much money would be saved each year in producing the same electrical power: (a) if the thermal efficiency of a reactor could be increased from $\frac{1}{3}$ to 0.4? (b) if the fuel consumed were 2% enrichment rather than 3%?

13.3. (a) Taking account of Xe-135 production, absorption, *and decay*, show that the balance equation is

$$N_x(\phi\sigma_{ax} + \lambda_x) = \phi N_F \sigma_{fF} y.$$

(b) Calculate λ_x and the ratio of absorption rates in Xe-135 and fuel if ϕ is 2×10^{13} cm⁻²-sec.

13.4. Soon after a reactor starts operating, the fission product Xe-135 builds up to

†The metric ton (or tonne) is 1000 kg (2200 lb), slightly larger than the ordinary ton.

its essentially steady level. What amount of reduction in the boron content must be made to compensate for this new absorption?

13.5. The initial concentration of boron in a 10,000 ft^3 reactor coolant system is 1500 ppm (the number of micrograms of additive per gram of diluent). What volume of solution of concentration 8000 ppm should be added to achieve a new value of 1600 ppm?

13.6. An adjustment of boron content from 1500 to 1400 ppm is made in the reactor described in Problem 13.5. Pure water is pumped in and then mixed coolant and poison are pumped out in two separate steps. For how long should the 500 ft^3/min pump operate in each of the operations?

14

Energy Conversion Methods

Most of the energy released in fission or fusion appears as kinetic energy of a few high-speed particles. As these pass through matter, they slow down by multiple collisions and impart thermal energy to the medium. It is the purpose of this chapter to discuss the means by which this energy is transferred to a cooling agent and transported to devices that convert mechanical energy into electrical energy.

14.1 METHODS OF HEAT TRANSMISSION

We learned in basic science that heat, as one form of energy, is transmitted by three methods—conduction, convection, and radiation. The physical processes for each of these are different: In *conduction*, molecular motion in a substance at a point where the temperature is high causes motion of adjacent molecules, and a flow of energy toward a region of low temperature takes place. The rate of flow is proportional to the slope of the temperature, i.e., the temperature gradient. In *convection*, molecules of a cooling agent such as air or water strike a heated surface, gain energy, and return to raise the temperature of the coolant. The rate of heat removal is proportional to the difference between the surface temperature and that of the surrounding medium, and also dependent on the amount of circulation of the coolant in the vicinity of the surface. In *radiation*, molecules of a heated body emit and receive electromagnetic radiations, with a net transfer of energy that depends on the temperatures of the body and the adjacent regions, specifically on the difference between the temperatures raised to the fourth power. For reactors, this mode of heat transfer is generally of less importance than the other two.

14.2 CONDUCTION IN REACTOR FUEL

The transfer of heat by conduction in a flat plate (insulated on its edges) is reviewed. If the plate has a thickness x and cross-sectional area A, and the temperature difference between its faces is ΔT, the amount of heat Q that flows in a time t through the plate is given by the relation

$$Q = kAt\,\frac{\Delta T}{x},$$

where k is the conductivity, with typical units joules/sec-°C-cm. For the plate, the slope of the temperature is the same everywhere. In a more general case, the slope may vary with position, and the heat flow per unit area Q/At is proportional to the slope or gradient written as $\Delta T/\Delta x$.

This idea is applied to the conduction in a single fuel rod of a reactor, with the rate of supply of thermal energy by fission taken to be uniform throughout the rod. If the rod is long in comparison with its radius R, or if it is composed of a stack of pellets, most of the heat flow is in the radial direction. If the surface is maintained at a temperature T_s by the flow of coolant, the center of the rod must be at some higher temperature T_0. As expected, the temperature difference is large if the rate of heat generation per unit volume q or the rate of heat generation per unit length $q_1 = \pi R^2 q$ is large. We can show† that

$$T_0 - T_s = \frac{q_1}{4\pi k},$$

and that the temperature T is in the shape of a parabola within the rod. Figure 14.1 shows the temperature distribution.

Let us calculate the temperature difference $T_0 - T_s$ for a reactor fuel rod of radius 0.5 cm, at a point where the power density is $q = 200 \text{ W/cm}^3$. This corresponds to a linear heat rate $q_1 = \pi R^2 q = \pi(0.25)(200) = 157 \text{ W/cm}$ (or 4.8 kW/ft). Letting the conductivity of UO_2 be $k = 0.062 \text{ W/cm-°C}$, we find $T_0 - T_s = 200°C$ (or 392°F). If we wish to keep the temperature low along the center line of the fuel, to avoid structural changes or melting, the conductivity k should be high, the rod size small, or the reactor power level low.

---

†The amount of energy supplied within a region of radius r must flow out across the boundary. For a unit length of rod with volume πr^2 and surface area $2\pi r$, the generation rate is $\pi r^2 q$, equal to the flow rate $[-k(dT/dr)]2\pi r$. Integrating from $r = 0$, where $T = T_0$, we have $T = T_0 - (qr^2/4k)$. At the surface $T_s = T_0 - (qR^2/4k)$.

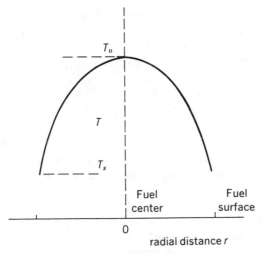

Fig. 14.1.   Temperature in fuel.

14.3 HEAT REMOVAL BY COOLANT

Convective cooling depends on many factors such as the fluid speed, the size and shape of the flow passage, and the thermal properties of the coolant, as well as on the area exposed and the temperature difference between surface and coolant $T_s - T_c$. Experimental measurements yield the "heat transfer coefficient" h, appearing in a working formula for the rate of heat transfer Q across a surface S,

$$Q = hS(T_s - T_c).$$

The units of h are typically W/cm^2-°C. In order to keep the surface temperature low, to avoid melting of the metal cladding of the fuel or to avoid boiling if the coolant is a liquid, a large surface area is needed or the heat transfer coefficient must be large, a low-viscosity coolant of good thermal conductivity is required, and the flow speed must be high.

As coolant flows along the many channels surrounding fuel pins in a reactor, it absorbs thermal energy and rises in temperature. Since it is the reactor power that is being extracted, we may apply the principle of conservation of energy. If the coolant of specific heat c enters the reactor at temperature T_c (in) and leaves at T_c (out), with a mass flow rate M, then the reactor thermal power P is

$$P = cM[T_c(\text{out}) - T_c(\text{in})] = cM\,\Delta T.$$

For example, let us find the amount of circulating water flow required to cool a reactor that produces 3000 MW of thermal energy. If the water enters at 300°C and leaves at 325°C, and the specific heat of water is 1 cal/g-°C or 4185 J/kg-°C, the mass flow rate is

$$M = \frac{P}{c\,\Delta T} = \frac{3000 \times 10^6 \text{ W}}{\left(4185 \dfrac{\text{W-sec}}{\text{kg-°C}}\right)(25°C)} = 29{,}000 \frac{\text{kg}}{\text{sec}}.$$

Noting that 1 gallon is 3.785 kg, this corresponds to 460,000 gal/min. To appreciate the magnitude of this flow, we can compare it with that from a garden hose of about 10 gal/min. The water for cooling a reactor is not wasted, of course, because it is circulated in a closed loop.

The temperature of coolant as it moves along any channel of the reactor can also be found by application of the above relation. In general, the power produced per unit length of fuel rod varies with position in the reactor because of the variation in neutron flux shape. For the special case of a *uniform* power along the z-axis with origin at the bottom (see Fig. 14.2a), the power per unit length is $P_1 = P/H$, where H is the length

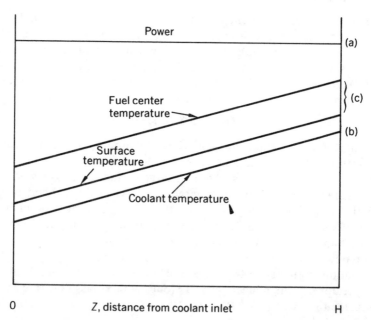

Fig. 14.2.   Temperature distributions along axis of reactor with uniform power.

of fuel rod. The temperature rise of coolant at z with channel mass flow rate M is then

$$T_c(z) = T_c(\text{in}) + \frac{P_1 z}{cM},$$

which shows that the temperature increases linearly with distance along the channel (see Fig. 14.2b). The temperature difference between coolant and fuel surface is the same at all points along the channel for this power distribution, and the temperature difference between the fuel center and fuel surface is also uniform. We can plot these as in Fig. 14.2c. The highest temperatures in this case are at the end of the reactor.

If instead, the axial power were shaped as a sine function (see Fig. 14.3a) with $P \sim \sin(\pi z/H)$, the application of the relations for conduction and convection yield temperature curves as sketched in Fig. 14.3b. For

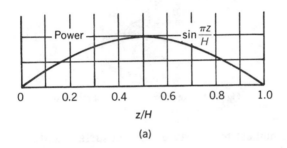

(a)

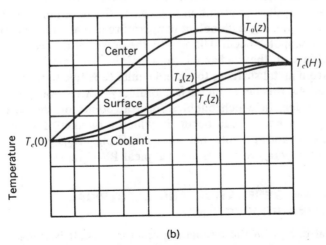

(b)

Fig. 14.3. Temperature distributions along channel with sine function power.

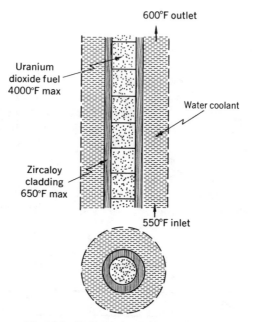

Fig. 14.4.   Reactor channel heat removal.

this case, the highest temperatures of fuel surface and fuel center occur between the halfway point and the end of the reactor. In the design of a reactor, a great deal of attention is given to the determination of which channels have the highest coolant temperature and at which points on the fuel pins "hot spots" occur. Ultimately, the power of the reactor is limited by conditions at these channels and points. The mechanism of heat transfer from metal surfaces to water is quite sensitive to the temperature difference. As the latter increases, ordinary convection gives way to *nucleate boiling,* in which bubbles form at points on the surface, and eventually *film boiling* can occur, in which a blanket of vapor reduces heat transfer and permits hazardous melting. Figure 14.4 indicates maximum temperature values for a typical PWR reactor.

14.4 STEAM GENERATION AND ELECTRICAL POWER PRODUCTION

Thermal energy in the circulating reactor coolant is transferred to a working fluid such as steam, by means of a heat exchanger or steam generator. In simplest construction, this device consists of a vessel partly

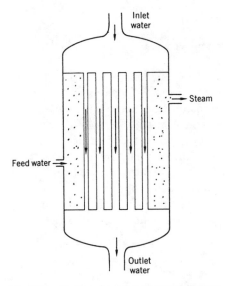

Fig. 14.5. Heat exchanger or steam generator.

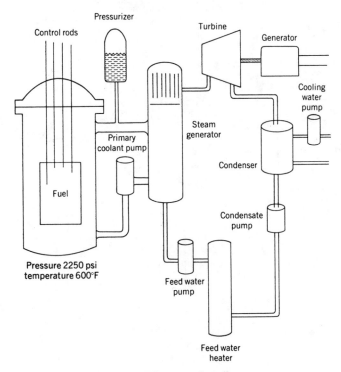

Fig. 14.6. PWR system flow diagram.

filled with water, through which many tubes containing heated water from the reactor pass, as in Fig. 14.5. Steam is evolved and flows to a turbine, while the water returns to the reactor. The conversion of thermal energy of steam into mechanical energy of rotation of a turbine and then to electrical energy from a generator is achieved by conventional means. Steam at high pressure and temperature is allowed to strike the blades of a turbine, which drives the generator. The exhaust steam is passed through a heat exchanger that serves as condenser, and the condensate is returned to the steam generator as feed water. Cooling water for the condenser is pumped from a nearby river, lake, or pond, which eventually receives the

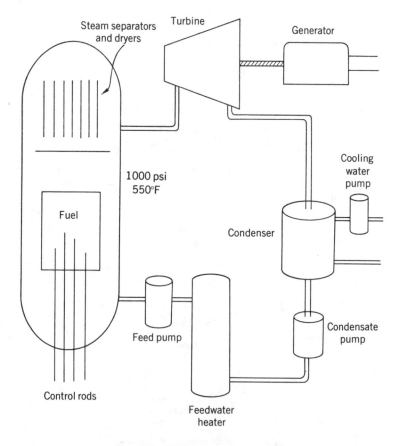

Fig. 14.7. BWR system flow diagram.

Fig. 14.8. A nuclear power plant. (Courtesy of Yankee Atomic Electric Company and the United States Atomic Energy Commission.)

waste heat from the energy conversion process. In some installations, a special cooling tower is employed to discharge waste heat.

Figures 14.6 and 14.7 show the flow diagrams for the reactor systems of the PWR and BWR type. In the PWR, a pressurizer maintains the pressure in the system at the desired value. It uses a combination of immersion electric heaters and water spray system to control the pressure. Figure 14.8 shows the Yankee PWR nuclear power plant, at Rowe, Massachusetts, in operation since 1961.

14.5 SUMMARY

The principal modes by which fission energy is transferred in a reactor are conduction and convection. The radial temperature in a fuel pellet is approximately parabolic. The rate of heat transfer from fuel surface to coolant by convection is directly proportional to the temperature difference. The allowed power level of a reactor is governed by the temperatures at local "hot spots." Coolant flow along channels extracts thermal energy and delivers it to an external circuit consisting of a heat exchanger (PWR), a steam turbine that drives an electrical generator, a steam condenser, and various pumps.

14.6 PROBLEMS

14.1 Show that the temperature varies with radial distance in a fuel pin of radius R according to

$$T(r) = T_s + (T_0 - T_s)[1 - (r/R)^2],$$

where the center and surface temperatures are T_0 and T_s, respectively. Verify that the formula gives the correct results at $r = 0$ and $r = R$.

14.2. Explain the advantage of a circulating fuel reactor, in which fuel is dissolved in the coolant. What disadvantages are there?

14.3. If the power density of a uranium oxide fuel pin, of radius 0.6 cm, is 500 W/cm^3, what is the rate of energy transfer per centimeter across the fuel pin surface? If the temperatures of pin surface and coolant are 300°C and 250°C, what must the heat transfer coefficient h be?

14.4. A reactor operates at thermal power of 2500 MW, with water coolant mass flow rate of 15,000 kg/sec. If the coolant inlet temperature is 275°C, what is the outlet temperature?

15

Breeder Reactors

The most important feature of the fission process is, of course, the enormous energy release from each reaction. Another significant fact, however, is that for each neutron absorbed in a fuel such as U-235, more than two neutrons are released. In order to maintain the chain reaction, only one is needed. Any extra neutrons available can thus be used to produce other fissile materials such as Pu-239 and U-233 from the "fertile" materials, U-238 and Th-232, respectively. The nuclear reactions yielding the new isotopes were described in Chapter 7. If losses of neutrons can be reduced enough, the possibility exists for new fuel to be generated in quantities as large or even larger than the amount consumed, a situation called "breeding."

In this chapter we shall (a) examine the relationship between the reproduction factor and breeding, (b) describe the physical features of the liquid metal fast breeder reactor, and (c) look into the compatibility of uranium fuel resources and requirements.

15.1 THE CONCEPT OF BREEDING

The ability to convert significant quantities of fertile materials into useful fissile materials depends crucially on the magnitude of the reproduction factor, η, which is the number of neutrons produced per neutron absorbed in fuel. If ν neutrons are produced per fission, and the ratio of fission to absorption in fuel is σ_f/σ_a, then the number of neutrons per absorption is

$$\eta = \frac{\sigma_f}{\sigma_a} \, \nu.$$

The greater its excess above 2, the more likely is breeding. It is found that both ν and the ratio σ_f/σ_a increase with neutron energy and thus η is larger for fast reactors than for thermal reactors. Table 15.1 compares values of η for the main fissile isotopes in the two widely differing neutron energy ranges designated as thermal and fast. Inspection of the table shows that it is more difficult to achieve breeding with U-235 and Pu-239 in a thermal reactor, since the 0.08 or 0.12 neutrons are very likely to be lost by absorption in structural materials, moderator, and fission product poisons. A thermal reactor using U-233 is a good prospect, but the fast reactor using Pu-239 is the most promising candidate for breeding.

Table 15.1. Values of Reproduction Factor η.

	Neutron energy	
Isotope	Thermal	Fast
U-235	2.08	2.3
Pu-239	2.12	2.7
U-233	2.28	2.45

Absorption of neutrons in Pu-239 consists of both fission and capture, the latter resulting in the isotope Pu-240. If it captures a neutron, the fissile isotope Pu-241 is produced.

The ability to convert fertile isotopes into fissile isotopes can be measured by the *conversion ratio* (CR), which is defined as

$$CR = \frac{\text{fissile atoms produced}}{\text{fissile atoms consumed}}.$$

The fissile atoms are produced by absorption in fertile atoms; the consumption includes fission and capture. Examination of the neutron cycle for a thermal reactor (Fig. 12.4) shows that the conversion ratio is dependent on η_F, the reproduction factor of the *fissile* material used, on ϵ, the fast fission factor, and on the amount of neutron loss by leakage and by absorption in nonfuel material, the sum of which is represented by a factor l. At the beginning of operation of a reactor, the conversion ratio is given by

$$CR = \eta_F \epsilon - 1 - l.$$

We can compare values of CR for different systems. For example, in a thermal reactor with $\eta_F = 2.08$, $\epsilon = 1.05$, and $l = 0.68$, the conversion ratio is 0.50. By adopting a reactor concept for which η_F is larger and by reducing the amount of neutron leakage, a conversion ratio of 1 can be obtained, which means that a new fissile atom is produced for each one consumed. Further improvements yield a conversion ratio greater than 1. For example, if $\eta_F = 2.7$, $\epsilon = 1$, and $l = 0.3$, the conversion ratio would be 1.4, meaning that 40% more fuel is produced than is used. Thus the value of CR ranges from zero in the pure "burner" reactor, containing no fertile material, to numbers in the range 0.5–0.7 in typical "converter" reactors, to 1.0 or larger in the "breeder."

If unlimited supplies of uranium were available at very small cost, there would be no particular advantage in seeking to improve conversion ratios. One would merely burn out the U-235 in a thermal reactor, and discard the remaining U-238 or use it for non-nuclear purposes. Since the cost of uranium goes up as the accessible reserves decline, it is very desirable to use the U-238 atoms which comprise 99.28% of the natural uranium, as well as the U-235, which is only 0.72%. Similarly, the exploitation of thorium reserves is highly worthwhile.

We can gain an appreciation of how large the conversion ratio must be to achieve a significant improvement in the degree of utilization of uranium resources, by the following logic: Suppose that U-235 and Pu-239 are equally effective in multiplication in a thermal converter reactor, i.e., their η values are assumed to be comparable. Let us abbreviate CR by writing it as simply R. Now the consumption of one U-235 atom will use up R atoms of U-238 and yield R atoms of Pu-239. The latter are then returned to the reactor and burned to convert R^2 atoms. The next step gives R^3, and so on. The total number of U-238 atoms converted per U-235 atom consumed is then

$$R + R^2 + R^3 + \cdots = R(1 + R + R^2 + \cdots) = \frac{R}{1 - R}$$

(where we have used the formula for the geometric series, $R < 1$). If $R = 0.6$ for example,
$$\frac{R}{1 - R} = \frac{0.6}{0.4} = 1.5.$$

This is far from complete conversion of the U-238. If all of the 0.72% U-235 in natural uranium feed were burned, the amount of U-238 converted would be only $(1.5)(0.72) = 1.08\%$, leaving about 99% unused. The percentage of the original *uranium* used is only $0.72 + 1.08 = 1.80$, or less than 2%. We can find what R must be to achieve complete conversion. If all of the atoms in the 99.28% that is U-238 are converted while all of the atoms in the 0.72% that is U-235 are consumed it is necessary that

$$\frac{R}{1-R} = \frac{99.28}{0.72}.$$

Solving, R = 0.9928, which is very close to unity. When one considers the effect of inevitable losses of uranium in reprocessing and refabrication, it is found that for practical purposes, the conversion ratio must be well above 1 in order to use all of the U-238.

When the conversion ratio is larger than 1, as in a fast breeder reactor, it is instead called the breeding ratio (BR), and the breeding gain (BG) = BR − 1 represents the extra plutonium produced per atom burned. The doubling time (DT) is the length of time required to accumulate a mass of plutonium equal to that in a reactor system, and thus provide fuel for a new breeder. The smaller the inventory of plutonium in the cycle and the larger the breeding gain, the quicker will doubling be accomplished. The technical term "specific inventory" is introduced, as the ratio of plutonium mass in the system to the electrical power output. Values of this quantity of 2.5 kg/MWe are sought. At the same time, a very long fuel exposure is desirable, e.g., 100,000 MWd/tonne, in order to reduce fuel fabrication costs. A breeding gain of 0.4 would be regarded as excellent, but a gain of only 0.2 would be very acceptable.

15.2 THE FAST BREEDER REACTOR

Several fast reactors have been built and tested, and the feasibility of achieving a breeding ratio that is greater than 1 has been established. The next step is the demonstration that a practical large-scale fast breeder power reactor can be built. The United States Atomic Energy Commission, in cooperation with the nuclear industry, has under way a high-priority national research and development program, with the goal of achieving a liquid metal fast breeder reactor (LMFBR) for commercial power production in the early 1980s. A "demonstration" plant in the range 350–400 MWe is being constructed, with a target date of 1979 for operation.

The use of liquid sodium Na-23 as coolant assures that there is little neutron moderation in the fast reactor. The element sodium melts at 208°F, boils at 1618°F, and has excellent heat transfer properties. With such a high melting point, pipes containing sodium must be thermally insulated and heated electrically to prevent freezing. The coolant be-

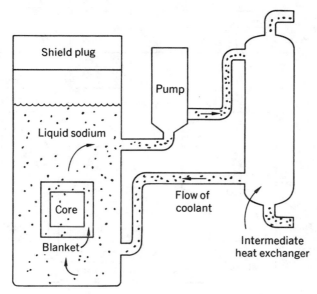

Fig. 15.1.   Loop system for LMFBR.

comes radioactive as the result of neutron absorption, producing the 15-hr Na-24. Great care must be taken to prevent contact between sodium and water or air, which would result in a serious fire, accompanied by the spread of radioactivity. To avoid such an event, an intermediate heat exchanger is employed. Heat is transferred in it from radioactive sodium to nonradioactive sodium.

Two physical arrangements of the reactor core, pumps, and heat exchanger are possible, shown schematically in Figs. 15.1 and 15.2. The "loop" type is similar to the thermal reactor system, while in the "pot" type all of the components are immersed in a pool of liquid sodium. There are advantages and disadvantages to each concept, but both appear practical. In the demonstration plant the loop is used.

The fuel for the LMFBR is in the form of cylindrical pellets, initially as a mixture of oxide UO_2 and PuO_2, inserted in stainless steel tubes. A bundle of around 100 pins is formed into a hexagonal fuel element, and 348 such elements are assembled into a core, as sketched in Fig. 15.3. A "blanket" of natural or depleted uranium surrounds the core to absorb neutrons that would otherwise be lost by leakage. The plutonium produced in the blanket is recovered by reprocessing. Some of the other design features of the demonstration reactor, to be located near Oak

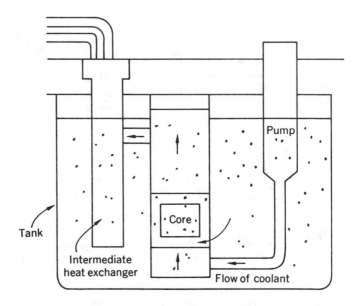

Fig. 15.2. Pot system for LMFBR.

Ridge, Tennessee, on a TVA site, are listed below:

thermal power	1000 MW
efficiency	36%
plutonium mass	1109 kg
sodium temperatures	
inlet	740°F
outlet	1000°F
steam conditions	
temperature	900°F
pressure	1450 psi

Although principal attention is being given in the United States to the fast breeder, there remains some interest in thermal breeding using the thorium, uranium-233 cycle. Concepts include (a) uranium and thorium fuel particles suspended in heavy water, (b) fuel and fertile elements as fluoride compounds mixed with other salts in molten form, and (c) a high-temperature gas-cooled graphite moderated reactor containing also a compound of beryllium, in which the $(n, 2n)$ reaction occurs.

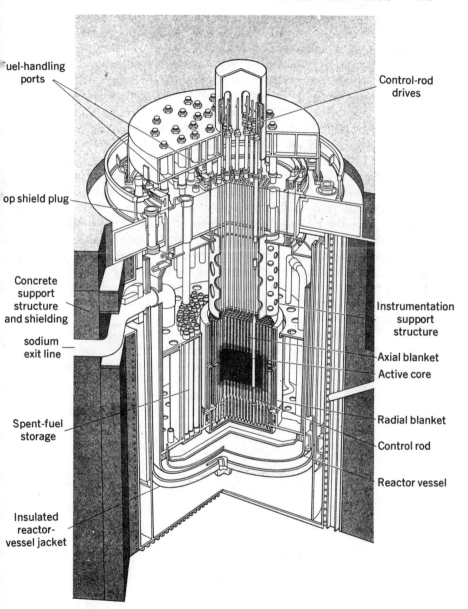

Fuel-handling ports

Control-rod drives

Top shield plug

Concrete support structure and shielding

Instrumentation support structure

sodium exit line

Axial blanket

Active core

Spent-fuel storage

Radial blanket

Control rod

Reactor vessel

Insulated reactor-vessel jacket

Fig. 15.3. Liquid metal fast breeder reactor. (From "Fast Breeder Reactors" by Glenn T. Seaborg and Justin L. Bloom. Copyright © November 1970 by Scientific American, Inc. All rights reserved.)

15.3 BREEDING AND URANIUM RESOURCES

The urgency of establishing technical feasibility and commercial viability of the breeder can be seen by a comparison of uranium reserves and uranium requirements. Most of the uranium deposits in the United States are in sandstones of New Mexico and Wyoming, with lesser amounts in Texas, Utah, and Colorado. The uranium ore reserves at price no higher than $8 per pound of U_3O_8 amount to 273,000 tons and potential resources are estimated to be 450,000 tons.[†] This total of 723,000 tons is to be compared with the expected demand of some 1,800,000 tons in the period 1973–1990, most of which is required to fuel conventional burners and converter reactors. When the low-cost fuel is exhausted, it will be necessary to turn to higher cost resources in the range $8–15 per pound U_3O_8, in amount available estimated to be more than a million tons. Before the end of the century if a breeder reactor were not developed, the demand for uranium by burner and converter reactors would inevitably push the price well above $15 per pound. On the other hand, a successful breeder reactor would make full use of the U-238 in natural uranium, and thus reduce the demand for uranium by a factor of around 50 for the same amount of power output.

The "reasonably assured" uranium reserves for a number of noncommunist countries are shown in Table 15.2. The estimates refer to U_3O_8 at less than $10 per pound. It is clear that importation of fuel is not a

Table 15.2. Uranium Reserves.[†]

Country	Tons U_3O_8
South Africa	263,000
Canada	241,000
Australia	140,000
Niger	52,000
France	47,500
Gabon	26,000
Others	67,500
Total	837,000

†From *Uranium '73*, Atomic Industrial Forum, New York, N.Y., 1974.

†*The Nuclear Industry 1973*, United States Atomic Energy Commission.

solution for the energy problem of the United States, since the United States has about a third of the uranium reserves, and energy needs of the rest of the world will absorb the rest. To meet the long-term world energy requirements, the controlled fusion process, described in the next chapter, will be needed.

15.4 SUMMARY

If the value of the neutron reproduction factor η is larger than 2 and losses of neutrons are minimized, breeding can be achieved in which more fuel is produced than is consumed. The conversion ratio (CR) measures the ability of a reactor system to transform a fertile isotope, e.g., U-238, into a fissile isotope, e.g., Pu-239. Complete conversion requires a value of CR of nearly 1. A fast breeder reactor using liquid sodium as coolant is being developed for commercial power. Estimated low-cost uranium reserves are sufficient only for a few decades of operation of reactors unless the breeder is successfully developed.

15.5 PROBLEMS

15.1. What are the largest conceivable values of the conversion ratio and the breeding gain?

15.2. An "advanced converter" reactor is proposed that will utilize 50% of the natural uranium supplied to it. Assuming all the U-235 is used, what must the conversion ratio be?

15.3. Explain why the use of a natural uranium "blanket" is an important feature of a breeder reactor.

15.4. Compute η and BG for a fast Pu-239 reactor if $\nu = 2.98$, $\sigma_f = 1.85$, $\sigma_c = 0.26$, and $l = 0.41$. (Note that the fast fission factor ϵ need not be included.)

15.5. With a breeding ratio BR = 1.20, how many kilograms of fuel will have to be burned in a fast breeder reactor operating only on plutonium in order to accumulate an extra 1260 kg of fissile material? If the power of the reactor is 1250 MWt, how long will it take in days and years, noting that it requires approximately 1.3 g of plutonium per MWd?

15.6. Verify the relation for conversion ratio CR $= \eta_F \epsilon - 1 - l$ by study of the thermal reactor neutron cycle (Chapter 12), noting that $l = l_f + l_t + a_t$, where the three terms on the right are the amounts of fast leakage, thermal leakage, and absorption in nonfuel materials per neutron absorbed in U-235. Note that $k = 1$ for a critical reactor.

16

Fusion Reactors

A device that permits the controlled release of fusion energy is designated as a fusion reactor, in contrast with one yielding fission energy, the fission reactor. As discussed in Chapter 8, the potentially available energy from the fusion process is enormous. The possibility of achieving controlled thermonuclear power on a practical basis has not yet been demonstrated, but progress in recent years gives encouragement that fusion reactors will be in operation early in the twenty-first century. In this chapter we shall review the choices of nuclear reaction, study the requirements for feasibility and practicality, and describe the physical features of machines that have been tested.

16.1 COMPARISON OF FUSION REACTIONS

The main nuclear reactions that combine light isotopes to release energy, as described in Chapter 8, are the D–D, D–T, and D–^{3}He. There are advantages and disadvantages of each. The reaction involving only deuterium uses an abundant natural fuel, available from water by isotope separation. However, the energy yield from the two equally likely reactions is low (4.03 and 3.27 MeV). Also, the cross section for the D–D reaction is small compared with that for the D–T interaction (see Fig. 16.1), thus requiring a higher plasma temperature. Despite the disadvantages, the D–D reaction will probably be the eventual long-range device, on the basis of fuel requirements.

The D–T reaction is recalled as

$$^2_1H + ^3_1H \rightarrow ^4_2He + ^1_0n + 17.6 \text{ MeV}.$$

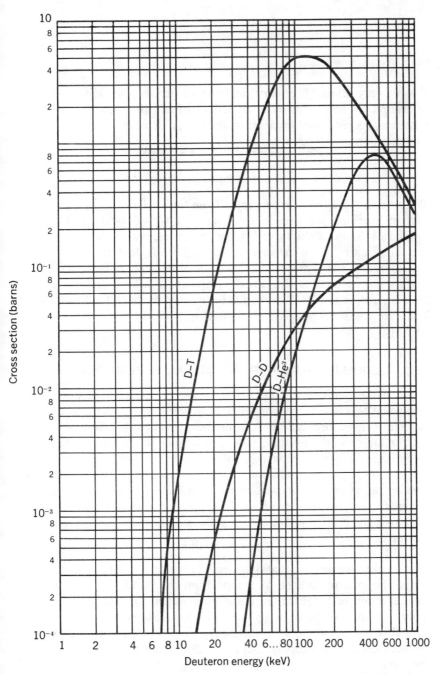

Fig. 16.1. Cross sections for fusion reactions. (From *Controlled Thermonuclear Reactions* by Glasstone/Lovberg © 1960 by Litton Educational Publishing, Inc. Reprinted by permission of Van Nostrand–Reinhold Company.)

It has a very favorable energy yield and a high cross section, with a peak at around 100 keV. However, an artificial fuel, tritium, is required. The latter can be generated by neutron absorption in lithium, according to the two reactions

$$^6_3\text{Li} + ^1_0\text{n} \rightarrow ^3_1\text{H} + ^4_2\text{He} + 4.8 \text{ MeV},$$

$$^7_3\text{Li} + ^1_0\text{n} \rightarrow ^3_1\text{H} + ^4_2\text{He} + ^1_0\text{n} - 2.5 \text{ MeV}.$$

We note that the neutron can come from the D–T reaction, suggesting a sort of breeding process. In the long run, use of the D–T reaction is limited by the availability of lithium resources, which though abundant are not nearly as inexhaustible as those of deuterium. All things considered, the D–T fusion reactor is the most likely to be operated first, and its success may lead to the development of a D–D reactor. There is little interest in the D–^{3}He reaction because of the need for a rare isotope as target and the low cross section, in spite of the high energy yield of 18.3 MeV.

16.2 REQUIREMENTS FOR PRACTICAL FUSION REACTORS

Since the purpose of any fusion device is to generate power, it is important to know the dependence of power density on factors such as plasma temperature and particle number densities. We first concentrate on conditions in the plasma, without reference to the surroundings. Our experience with the calculation of reaction rates (Chapter 5) may be applied. In a plasma containing n_D deuterons and n_T tritons per cubic centimeter, the particles interact with an effective average of the product of cross section and speed σv. The reaction rate per unit volume is $n_D n_T \sigma v$, and if the sensible energy yield is E for each reaction, the fusion power density is

$$p = n_D n_T \sigma v E.$$

We should assume here that the energy E is that of the helium atom only, since the neutron is free to escape from the plasma. On the basis of conservation of kinetic energy and momentum, the $^4_2$He atom will have only $\frac{1}{5}$ of the reaction energy, i.e., $17.6/5 = 3.5$ MeV. When the plasma contains equal numbers of the heavy particles, $n_D = n_T = n$, then the power density p is proportional to n^2, the square of the particle number density.

The radiation losses described in Chapter 8 pose a serious limitation on the achievement of practical fusion power. The radiated power density p_r exceeds that from fusion, p, until the ignition temperature is reached.

Ideally, we should like to maintain a steady electrical discharge with constant net power output. However, it will be sufficient if the reaction could be repeated periodically in short bursts of time of duration τ. If the frequency is high enough, the average power will be adequate. For a given value of τ and operating temperature T there must be a certain particle number density n for this mode of operation to be meaningful, according to the *Dawson criterion*

$$n\tau = \text{constant},$$

where the constant depends on T. The origin of this simple rule of thumb is as follows: We assume that energy must be provided to bring the plasma up to a temperature T where the fusion reaction is favorable. In one cubic centimeter there are $2n$ nuclei, each brought to average energy $\frac{3}{2} kT$, requiring an energy addition of $3nkT$. (We ignore electron energy in comparison with ion energy.) It is also necessary to supply energy to compensate for the radiation loss, as the product of power density p_r and time of pulse τ. Thus the total supplied must be $3nkT + p_r\tau$. After a pulse, we have the sum of the thermal energy of the plasma, the fusion energy, and the radiation energy, but with an efficiency of recovery ϵ that is less than 1, the available energy is only $\epsilon(3nkT + p\tau + p_r\tau)$. Equating the energies and noting that each power is proportional to n^2, we can solve

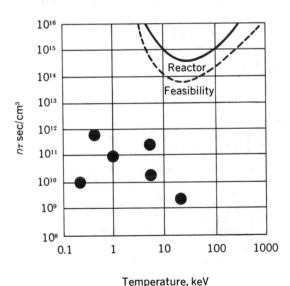

Fig. 16.2.   Status of fusion.

for the product $n\tau$ as a function of T. The break-even point for the D–T reaction is around $n\tau = 10^{14}$ sec/cm^3. However, a fusion reactor with no energy extraction is useless, and it is necessary for $n\tau$ to be at least 10 times higher. The goal of 10^{15} sec/cm^3 can be reached with many different combinations of n and τ, e.g., $n = 10^{15}$, $\tau = 1$ sec or $n = 10^{21}$, $\tau = 10^{-6}$ sec. Figure 16.2 shows the accomplishments of some of the machines being tested as points on a graph of $n\tau$ against T, and shows the range of conditions that must be reached for feasibility of controlled fusion, and the more difficult goal of a practical fusion reactor.

16.3   FUSION DEVICES

Many complex machines have been devised to generate a plasma and to provide the necessary electric and magnetic fields to achieve confinement of the discharge. We shall examine a few of these to illustrate the variety of possible approaches.

First, consider a simple discharge tube consisting of a gas-filled glass cylinder with two electrodes as in Fig. 16.3a. This is similar to the familiar fluorescent lightbulb. Electrons accelerated by the potential difference cause excitation and ionization of atoms. The ion density and temperature of the plasma that is established are many orders of magnitude below that needed for fusion. To reduce the tendency for charges to diffuse to the walls and be lost, a current-carrying coil can be wrapped around the tube, as sketched in Fig. 16.3b. This produces a magnetic field directed along the axis of the tube, and charges move in paths described by a helix, the

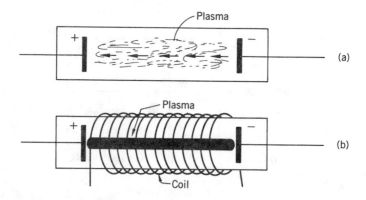

Fig. 16.3.   Electrical discharges: (a) without magnetic field, (b) with magnetic field.

shape of a stretched coil spring. The motion is quite similar to that of ions in the cyclotron (Chapter 9) or the mass spectrograph (Chapter 10). The radii in typical magnetic fields and plasma temperatures are the order of 0.1 mm for electrons and near 1 cm for heavy ions (see Problem 16.1). In order to further improve charge density and stability, the current along the tube is increased to take advantage of the pinch effect, a phenomenon related to the electromagnetic attraction of two wires that carry current in the same direction. Each of the charges that move along the length of the tube constitutes a tiny current, and the mutual attractions provide a constriction in the discharge.

Neither of the above magnetic effects prevent charges from moving freely along the discharge tube, and losses of both ions and electrons are experienced at the ends. Two solutions of this problem have been tried. One is to wrap extra current-carrying coils around the tube near the ends (see Fig. 16.4a), increasing the magnetic field there. Figures 16.4b and

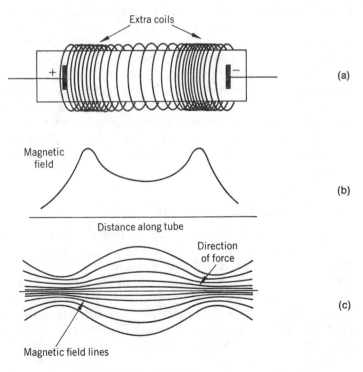

Fig. 16.4.   Magnetic mirror.

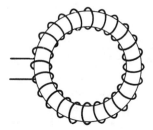

Fig. 16.5.   Plasma confinement in torus.

16.4c show the field variation and the shape of the field lines. There is a tendency for charges to be forced back into the region of weak field, i.e., to be reflected. Such an arrangement is thus called a "mirror" machine, but it is by no means perfectly reflecting.

Another solution is to produce the discharge in a doughnut-shaped tube (torus), as shown in Fig. 16.5. Since the tube has no ends, the magnetic field produced by the coils is continuous, and the free motion of charges along the field lines does not result in any losses. Without any other forces acting, the variation in magnetic field in the radial direction in the tube tends to allow charge diffusion to the walls. The introduction of a current through the plasma along the axis of the torus has some important effects. It provides a magnetic field in the form of circles around the current, that tends to "pinch" the plasma and prevent its dispersal. It also serves as the means for heating the plasma to the ignition temperature. The rate of energy deposition by a current I through a medium with electrical resistance R is given by I^2R. The resistance is principally due to the electron-ion collisions.

16.4   EXTRACTION OF ENERGY

The achievement of a self-sustaining fusion reaction in a plasma must be accompanied by the development of a practical method of extracting thermal energy. Let us visualize a system that will reveal some of the scientific and engineering problems to be solved. The reactor will probably consist of a long tube or circular ring, with many coaxial regions, as sketched in Fig. 16.6. In the highly ionized plasma, the D-T reaction produces alpha particles and neutrons. The kinetic energies of the alpha particles would be transferred to the plasma and ultimately extracted. Neutrons would be slowed down and absorbed in a coolant such as liquid lithium that flows between two walls, extracting heat from the inner one, absorbing neutron energy, and delivering the total energy to an external

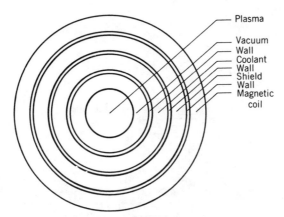

Fig. 16.6.   Hypothetical fusion reactor.

heat exchanger. The reaction of neutrons with lithium could give rise to the supply of tritium needed for the basic D–T reaction in the plasma. A shield is needed to prevent the magnetic coils from overheating from conducted heat and electromagnetic radiation from the plasma.

Also under study are two advanced concepts related to the fusion process. One involves the expansion of the plasma into a conical region with ion and electron collection electrodes in the walls (see Fig. 16.7).

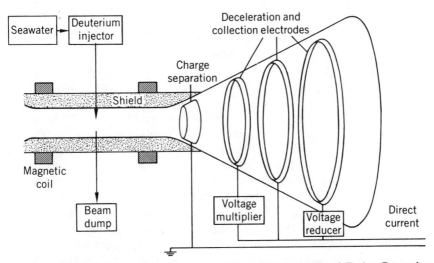

Fig. 16.7.   Direct conversion fusion reactor. (From *The Prospects of Fusion Power* by William C. Gough and Bernard J. Eastlund. Copyright © February 1971 by Scientific American, Inc. All rights reserved.)

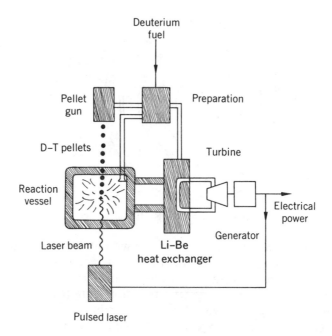

Fig. 16.8. Laser-pellet reactor.

This system could provide direct conversion of nuclear-thermal energy into electrical energy, and thus would be very efficient. Another consists of controlled explosions of pellets of deuterium and tritium. This process is not thermonuclear since the reaction does not take place in an ionic gas. As sketched in Fig. 16.8, pellets at very low temperatures are injected into a chamber and bombarded by a pulsed laser beam. This intense highly directed beam of light energy heats the D–T mixture above the ignition temperature and particle inertia provides the basis for confinement for the length of time for the fusion reaction to take place.

A novel idea for the utilization of thermonuclear plasmas is the "fusion torch," which could help solve the world's waste-disposal problem by making possible the recycling of products. Waste materials would be converted into their basic elements through exposure to a high-temperature plasma. The torch might also perform various chemical reactions by applying ultraviolet (UV) radiation from the discharge. Some of the applications of the fusion torch that have been suggested are listed.

Uses of Plasma Energy	*Uses of UV Radiation*
Ore processing	Desalination of sea water
Waste product treatment	Radiative heating
Alloy separation	Waste sterilization
Destruction of toxic chemicals	Cultivation of algae for food
Disposal of plastics	Production of ozone

16.5 SUMMARY

A fusion reactor, yet to be developed, would provide power from a controlled fusion reaction. Of the several possible nuclear reactions, the one that will probably be used first involves deuterium and tritium (produced by neutron absorption in lithium). A D–T reactor that yields significantly more energy than is supplied to achieve the reaction must have a product $n\tau$ above 10^{15}, where n is the particle number density and τ is the duration of pulsed operation. Many types of experimental machines have been tested. Most involve an electrical discharge (plasma) that is influenced by magnetic fields due to currents in external coils or within the discharge tube. A fusion reactor will probably consist of a doughnut-shaped tube, with several concentric layers. Advanced concepts include direct conversion, the laser-pellet method, and the fusion torch.

16.6 PROBLEMS

16.1. Noting that the radius of motion R of a particle of charge q and mass m in a magnetic field B is $R = mv/qB$ and that the kinetic energy of rotation in the x-y plane is $\frac{1}{2}mv^2 = kT$, find the radii of motion of electrons and deuterons if B is $10\,\text{Wb/m}^2$ and kT is $100\,\text{keV}$.

16.2. Show that the effective nuclear reaction for a fusion reactor using deuterium, tritium, and lithium-6 is

$$^2_1\text{H} + ^6_3\text{Li} \rightarrow 2\,^4_2\text{He} + 22.4\,\text{MeV}.$$

16.3. Verify the statement that in the D–T reaction the ^4_2He particle will have $\frac{1}{5}$ of the energy.

16.4. (a) Assuming that in the D–D fusion reaction the fuel consumption is $0.151\,\text{g/MWd}$ (Problem 8.3), find the energy release in J/kg. By how large a factor is the value larger or smaller than that for fission?
(b) If heavy water costs $30 per pound, what is the cost of deuterium per kilogram?
(c) Noting $1\,\text{kWh} = 3.6 \times 10^6\,\text{J}$, find from (a) and (b) the energy cost in mills/kWh.

Part III Nuclear Energy and Man

The discovery of nuclear reactions that yield energy, radiation, and radioisotopes is generally regarded as one of the most significant scientific contributions of the twentieth century, in that it showed the possibility of enormous human benefit or of world destruction. It is thus understandable that nuclear energy should be a controversial subject. Many people have deplored its initial use for military purposes, while others regarded the action as necessary under the existing circumstances. Some believe that the discovery of nuclear energy should somehow have been avoided, while others hold that the revelation of natural phenomena is inevitable. Many uninformed persons see no distinction between nuclear weapons and nuclear reactors, while others recognize that the two are entirely different. A small group of scientists would abandon the use of nuclear energy on the basis of risks, while many other knowledgeable persons believe that it should be applied to national and world energy needs.

The variety of viewpoints on nuclear energy is but a part of a larger picture—the recent growth in concern about the effects of science and technology, which are claimed by some to be the source of many of the problems of advanced countries. Such a reaction is natural when one learns the extent of waste release and the effect of environmental pollution on wildlife and human beings. There is no doubt that there exists a sequence of events starting with a scientific discovery, continuing with a technological and commercial application, and too often ending with a new environmental or social problem. It does not follow, however, that the investigations should not have been initiated, but rather that adequate information about possible side effects should have been developed and that positive recommendations to prevent harmful releases should have been made. Nor does it suggest that new beneficial technology should be discouraged, but that realistic appraisals of the costs of ensuring adequate protection be developed and that the public be made aware of the additional expense. Finally, excessive concern about the effect of the industrial byproducts is paradoxical in a world in which two-thirds of the population receives none of the benefits of health, freedom from drudgery, and high standard of living that we enjoy as fruits of a technological civilization.

Decisions as to the uses of science are subject to ethical and moral criteria, but science itself, as a process of investigation and a body of information that is developed, must be regarded as neutral. Every natural resource has mixed good and evil. For example, fire is most necessary and welcome for warmth of our homes and buildings but can devastate our forests. Water is required for survival of every living being but in the form of a flood can ruin our cities and land. Drugs can help cure diseases but can incapacitate or kill us. Explosives are valuable for mining and construction but are also a tool of warfare. So it is with nuclear energy—on one hand, we have the benefits of heat and radiation for many human needs; on the other, the possibility of bombs and radioactive fallout. The key to application for benefit or detriment lies in man's decisions, and the fear of evil uses should not deter us from taking full advantage of good uses.

In Part III we shall review the history of nuclear energy, examine its hazards and the means available for protection, and describe some of the many peaceful applications of nuclear energy to the betterment of mankind. Finally, we shall discuss the role of nuclear energy in the long-term survival of our species.

17

The History of Nuclear Energy

The development of nuclear energy exemplifies the consequences of scientific study, technological effort, and commercial application. We shall review the history for its relation to our cultural background, which should include man's endeavors in the broadest sense. The author subscribes to the traditional conviction that history is relevant. Present understanding is grounded in recorded experience, and while we cannot undo errors, we can avoid them in the future. We can hopefully establish concepts and principles about human attitudes and capability that are independent of time, to help guide future action. Finally, we can draw confidence and inspiration from the knowledge of what man has been able to accomplish.

17.1 THE RISE OF NUCLEAR PHYSICS

The science on which practical nuclear energy is based can be categorized as classical, evolving from studies in chemistry and physics for the last several centuries, and modern, that related to investigations over the last hundred years into the structure of the atom and nucleus. The modern era begins in 1879 with Crookes' achievement of ionization of a gas by an electric discharge. Thomson in 1897 identified the electron as the charged particle responsible for electricity. Roentgen in 1895 had discovered penetrating X-rays from a discharge tube, and Becquerel in 1896 found similar rays—now known as gamma rays—from an entirely different source, the element uranium, which exhibited the phenomenon of natural radioactivity. The Curies in 1898 isolated the radioactive element radium. As a part of his revolutionary theory of motion, Einstein in 1905 concluded that the mass of any object increased with its speed, and stated his now-famous formula $E = mc^2$, which expresses the equivalence of mass and energy. At that time, no experimental verification was available, and Einstein could not have foreseen the implications of his relation.

In the first third of the twentieth century, a host of experiments with the various particles coming from radioactive materials led to a rather clear understanding of the structure of the atom and its nucleus. It was learned from the work of Bohr and Rutherford that the electrically neutral atom is constructed from negative charge in the form of electrons surrounding a central positive nucleus, which contains most of the matter of the atom. Through further work by Rutherford in England around 1919, it was revealed that even though the nucleus is composed of particles bound together by forces of great strength, nuclear transmutations can be induced, e.g., the bombardment of nitrogen by helium yields oxygen and hydrogen.

In 1930, Bothe and Becker bombarded beryllium with alpha particles from polonium and found what they thought were gamma rays but which Chadwick in 1932 showed were neutrons. A similar reaction is now employed in a nuclear reactor to provide a source of neutrons. Artificial radioactivity was first reported in 1934 by Curie and Joliot. Alpha particles injected into nuclei of boron, magnesium, and aluminum gave new radioactive isotopes of several elements. The development of machines to accelerate charged particles to high speeds opened up new opportunities to study nuclear reactions. The cyclotron, developed in 1932 by Lawrence, was the first of a sequence of devices of ever-increasing capability.

17.2 THE DISCOVERY OF FISSION

During the 1930s, Enrico Fermi and his co-workers in Italy performed a number of experiments with the newly discovered neutron. He reasoned correctly that the lack of charge on the neutron would make it particularly effective in penetrating a nucleus. Among his observations was the neutron capture reaction and the efficiency for producing a great variety of radioisotopes by neutrons slowed in paraffin or water. Breit and Wigner provided the theoretical explanation of slow neutron processes in 1936. Fermi made measurements of the distribution of both fast and thermal neutrons and explained the behavior in terms of elastic scattering, chemical binding effects, and thermal motion in target molecules. During this period, many cross sections for neutron reactions were measured, including that of uranium, but the fission process was not identified.

It was not until January 1939 that Hahn and Strassmann of Germany reported that they had found the element barium as a product of neutron bombardment of uranium. Frisch and Meitner made the guess that fission was responsible for the appearance of an element that is only half as heavy as uranium. Fermi then suggested that neutrons might be emitted during the process, and the idea was born that a chain reaction that released great amounts of energy might be possible. The press picked up the idea, and many sensational articles were written. The information on fission, brought to the United States by Bohr on a visit from Denmark, prompted a flurry of activity at several universities, and by 1940 nearly a hundred papers had appeared in the technical literature. All of the qualitative characteristics of the chain reaction were soon learned—the moderation of neutrons by light elements, thermal and resonance capture, the existence of fission in U-235 by thermal neutrons, the large energy of fission fragments, the release of neutrons, and the possibility of producing transuranic elements, those beyond uranium in the periodic table.

17.3 THE DEVELOPMENT OF NUCLEAR WEAPONS

The discovery of fission, with the possibility of a chain reaction of explosive violence, was of especial importance at this particular time in history, since World War II had begun in 1939. Because of the military potential of the fission process, a voluntary censorship of publication on the subject was established by scientists in 1940. The studies that showed U-235 to be fissile suggested that the new element plutonium, discovered in 1941 by Seaborg, might also be fissile and thus also serve as a weapon material. As early as July 1939, four leading scientists—Szilard, Wigner, Sachs, and Einstein—had initiated a contact with President Roosevelt, explaining the possibility of an atomic bomb based on uranium. As a consequence a small grant of $6000 was made by the military to procure materials for experimental test of the chain reaction. (Before the end of World War II, a total of $2 billion had been spent, an almost inconceivable sum in those times.) After a series of studies, reports, and policy decisions, a major effort was mounted through the U.S. Army Corps of Engineers under General Groves. The code name "Manhattan District" (or "Project") was devised, with military security on all information.

Although a great deal was known about the individual nuclear reactions, there was great uncertainty as to the practical behavior: could a chain reaction be achieved at all? If so, could Pu-239 in adequate quantities be produced? Could a nuclear explosion be made to occur? Could U-235 be separated on a large scale? Attacks on these questions were initiated at several institutions, and design of production plants began almost concurrently, with great impetus provided by the involvement of the United States in World War II after the attack on Pearl Harbor in December 1941 by the Japanese. The distinct possibility that Germany was actively engaged in the development of an atomic weapon served as a strong stimulus to the work of American scientists, most of whom were in universities. They and their students dropped their normal work to enlist in some phase of the project.

The whole Manhattan Project consisted of parallel endeavors, with major effort in the United States and cooperation with the United Kingdom, Canada, and France. At the University of Chicago, tests preliminary to the construction of the first atomic pile were made; and on December 2, 1942, Fermi and his associates achieved the first chain reaction under the stands of Stagg Field. By 1944, the plutonium production reactors at Hanford, Washington had been put into operation, providing the new element in kilogram quantities. At the University of California at Berkeley, the electromagnetic separation "calutron" process for isolating U-235 was perfected, and government production plants at Oak Ridge, Tennessee were built in 1943. At Columbia University, the gaseous diffusion process for isotope separation was studied, forming the basis for the present production system, the first units of which were built at Oak Ridge. At Los Alamos, New Mexico, theory and experiment led to the development of the nuclear weapons, first tested at Alamogordo, New Mexico, on July 16, 1945, and later used at Hiroshima and Nagasaki in Japan.

The brevity of this account fails to reveal the dedication of scientists, engineers, and other workers to the accomplishment of national objectives, or the magnitude of the design and construction effort by American industry. Two questions are inevitably raised—Should the atom bomb have been developed? Should it have been used? Some of the scientists who worked on the Manhattan Project have expressed their feeling of guilt for having participated. Some insist that a lesser demonstration of the destructive power of the weapon should have been arranged, which would have been sufficient to end the conflict. Many others believed that the security of the United States was threatened and that the use of the weapon shortened World War II greatly and thus saved a large number of casualties on both sides. It is some comfort, albeit small, that the existence of nuclear weapons has served as a deterrent to a direct conflict between major powers for several decades.

The discovery of nuclear energy, with its tremendous potential for the betterment of mankind through new unlimited energy resources, and through radioisotopes and their radiation for research, medical diagnosis and treatment, and agricultural improvement can very well have benefits that far outweigh the detriments, particularly if we have sense enough not to use nuclear weapons again.

17.4 PEACEFUL APPLICATIONS OF NUCLEAR ENERGY

One of the first important events in the United States after World War II ended was the creation of the United States Atomic Energy Commission. This civilian federal agency was charged with the management and development of the nation's nuclear programs in behalf of peaceful applications. Several national laboratories were established to continue research in nuclear energy, at sites such as Oak Ridge, Argonne (near Chicago), Los Alamos, and Brookhaven (on Long Island). One of the main objectives of the AEC was to attack the problem of producing practical commercial nuclear power through research and development. This work was started at national laboratories, with Oak Ridge first looking at a gas-cooled reactor. They later planned a high-flux reactor fueled by highly enriched uranium alloyed with and clad with aluminum, and using water as moderator and coolant. The reactor was eventually built in Idaho as the Materials Testing Reactor. Before then, however, the idea arose of using a similar system for powering a nuclear submarine, and through the determination of H. G. Rickover, then a Navy captain, a development project was pushed through at Argonne. It was found that zirconium was preferable to aluminum because of its compatibility with high-temperature water under pressure. During the short period 1948 to 1953, many technical problems were resolved and a prototype submarine reactor was built and tested in Idaho. The Westinghouse Electric Corporation assisted in the development, did the design and construction, and later produced the reactor for installation in the first nuclear submarine, the *Nautilus*, which went to sea in 1955. This pressurized water reactor (PWR) concept was adapted by Westinghouse for use as the first commercial power plant at Shippingport, Pennsylvania, beginning operation in 1957 at an electric power output of 60 MW. Uranium dioxide pellets as fuel were first introduced in this design.

Two other reactor research and development programs were under way at Argonne over the same period. The first program was aimed at achieving power plus breeding of plutonium, using the fast reactor concept with liquid sodium coolant. The first electric power from a nuclear source was produced in late 1951 in the Experimental Breeder Reactor, and the possibility of breeding was demonstrated. This work has served as the basis for the present fast breeder reactor development program. The second program consisted of an investigation of the possibility of allowing water in a reactor to boil and generate steam directly. The principal concern was with the fluctuations and instability associated with the boiling. Tests called BORAX were performed that showed that a boiling reactor could operate safely, and work proceeded that led to electrical generation in 1955. The General Electric Company then proceeded to develop the boiling water reactor (BWR) concept further, with the first commercial reactor of this type put into operation at Dresden, Illinois in 1960.

On the basis of the initial successes of the PWR and BWR, and with the application of commercial design and construction know-how, Westinghouse and General Electric were able, in the early 1960s, to advertise large-scale nuclear plants of power around 500 MWe that would be competitive with fossil fuel plants in the cost of electricity. Immediately thereafter, there was a rapid move on the part of the electric utilities to order nuclear plants, and the growth in the late 1960s was phenomenal. Orders for nuclear steam supply systems for the years 1965–1970 inclusive amounted to around 88 thousand MWe, which was more than a third of all orders, including fossil fueled plants. The corresponding nuclear electric capacity was around a quarter of the total United States capacity at the end of that period of rapid growth.

In a parallel research and development program, the high-temperature gas-cooled reactor has been brought to commercial success by General Atomic, now a part of the Gulf Oil Corporation. The HTGR serves as an important alternative to light water reactors by virtue of its graphite moderator, helium coolant, and uranium–thorium fuel cycle. Adaptations of the concept to meet the goal of breeding are also promising.

Space does not permit the description of the important international aspects of the development and use of nuclear energy. It should be mentioned, however, that the United Kingdom as early as the 1950s developed and built many large graphite reactors for electrical power while Canada led the investigation of heavy water moderated reactors.

Both countries have continued to make important contributions to the technology. Several other European countries have strong nuclear programs, either independently, as in the case of France, or as a part of Euratom, a cooperative organization. The U.S.S.R. has a number of power reactors in use, and Japan is contributing strongly to the development of breeders. Valuable assistance to the nuclear field is provided by the International Atomic Energy Agency, based in Vienna.

Less than thirty years had elapsed between the discovery of the fission process and the advent of practical nuclear reactors for electrical power. The endeavor revealed a new concept—that large-scale national technological projects could be undertaken and successfully completed, by the application of large amounts of money and organization of the efforts of many sectors of society. The nuclear project in many ways served as a model for the United States space program of the 1960s. The most important lesson that the history of nuclear energy development may have for us is that urgent national and world problems can be solved by wisdom, dedication, and cooperation.

17.5 SUMMARY

A sequence of many investigations in atomic and nuclear physics spanning the period 1879–1939 led to the discovery of fission. New knowledge was developed about particles and rays, radioactivity, the equivalence of matter and energy, nuclear transmutation, and the structure of the atom and nucleus. The existence of fission suggested that a chain reaction involving neutrons was possible and that the process had military significance. A major national program was initiated that included uranium isotope separation by electromagnetic and gaseous diffusion, nuclear reactor studies, plutonium production, and weapons development, culminating in the use of the atomic bomb to end World War II. In the post-war period, emphasis was placed on peaceful applications of nuclear energy under the United States Atomic Energy Commission. Four reactor concepts—the pressurized water, boiling water, fast breeder, and gas cooled—evolved through work by national laboratories and industry. The first two were brought to commercial status in the 1960s. The nuclear power project of the United States demonstrated that major national goals could be achieved with sufficient effort. Many other countries also have active programs of nuclear power development.

18

Biological Effects of Radiation

All living species are exposed to a certain amount of natural radiation in the form of particles and rays. In addition to the sunlight, without which life would be impossible to sustain, all beings experience cosmic radiation from space outside the earth and natural background radiation from materials on the earth. There are rather large variations in the radiation from one place to another, depending on mineral content of the ground and on the elevation above sea level. Man and other species have survived and evolved within such an environment in spite of the fact that radiation has a damaging effect on biological tissue. The discovery by man of means to generate radiation, using X-ray machines, particle accelerators, or nuclear reactors, has added potential hazard to his existence. In assessing the importance of such man-made radiation, comparison is often made with levels in the naturally occurring background radiation.

We shall now describe the biological effect of radiation on cells, tissues, organs, and individuals, identify the units of measurement of radiation and its effect, and review the philosophy and practice of setting limits on exposure. Special attention will be given to regulations related to nuclear power plants.

A brief summary of modern biological information will be useful in understanding radiation effects. As we know, living beings consist of a great variety of species of plants and animals; they are all composed of cells, which carry on the processes necessary to survival. The simplest organisms such as algae and protozoa consist of only one cell, while complex beings such as man are composed of specialized organs and tissues that contain large numbers of cells, examples of which are nerve, muscle, epithelial, blood, skeletal, and connective. The principal components of a cell are the *nucleus* as control center, the *cytoplasm* containing vital substances, and the surrounding *membrane*, as a porous cell wall. Within the nucleus are the *chromosomes*, which are long threads containing hereditary material. The growth process involves a form of cell multiplication called *mitosis*—in which the chromosomes separate in order to form two new cells identical to the original one. The reproduction process involves a cell division process called *meiosis*—in which germ cells are produced with only half the necessary complement of chromosomes, such that the union of sperm and egg creates a complete new entity. The laws of heredity are based on this process. The genes are the distinct regions on the chromosomes that are responsible for inheritance of certain body characteristics. They are constructed of a universal molecule called DNA, a very long spiral staircase structure, with the stairsteps consisting of paired molecules of four types. Duplication of cells in complete detail involves the splitting of the DNA molecule along its length, followed by the accumulation of the necessary materials from the cell to form two new ones. In the case of man, there are 46 chromosomes, containing about four billion of the DNA molecule steps, in an order that describes each unique person.

18.1 PHYSIOLOGICAL EFFECTS

The various ways that moving particles and rays interact with matter discussed in earlier chapters can be reexamined in terms of biological effect. Our emphasis previously was on what happened to the radiation. Now, we are interested in the effects on the medium, which are viewed as "damage" in the sense that disruption of the original structure takes place, usually by *ionization.* We saw that energetic electrons and photons were capable of removing electrons from an atom to create ions; that heavy charged particles slowed down in matter by successive ionizing events; that fast neutrons in slowing imparted energy to target nuclei, which in turn serve as ionizing agents; that the loss of gamma ray may be accompanied by an electron–positron pair as new radiation; and that capture of a slow neutron results in a gamma ray and a new nucleus that recoils with appreciable energy.

As a good rule of thumb, 32 eV of energy is required on the average to create an ion pair. This figure is rather independent of the type of ionizing radiation, its energy, and the medium through which it passes. For instance, a single 4-MeV alpha particle would release about 10^5 ion pairs before stopping. Part of the energy goes into molecular excitation and the formation of new chemicals. Water in cells can be converted into free radicals such as H, OH, H_2O_2, and HO_2. Since the human body is largely water, much of the effect of radiation can be attributed to the chemical reactions of such products. In addition, direct damage can occur, in which the radiation strikes certain molecules of the cells, especially the DNA that controls all growth and reproduction.

The most important point from the biological standpoint is that the bombarding particles have energy, which can be transferred to atoms and molecules of living cells, with a disruptive effect on their normal function. Since an organism is composed of very many cells, tissues, and organs, a disturbance of one atom is likely to be imperceptible, but exposure to many particles or rays can alter the function of a group of cells and thus affect the whole system. It is usually assumed that damage is cumulative, even though some accommodation and repair surely takes place.

The physiological effects of radiation may be classified as *somatic*, which refers to the body and its state of health, and *genetic*, involving the genes that transmit hereditary characteristics. The somatic effects range from temporary skin reddening when the body surface is irradiated, to a life shortening of an exposed individual due to general impairment of the body functions, to the initiation of cancer in the form of tumors in certain organs or as the blood disease, leukemia. The term "radiation sickness" is loosely applied to the immediate effects of exposure to very large amounts of radiation. The genetic effect consists of mutations, in which progeny are significantly different in some respect from their parents, usually in ways that tend to reduce the chance of survival. The effect may extend over many generations.

Although the amount of ionization produced by radiation of a certain energy is rather constant, the biological effect varies greatly with the type of tissue involved. For radiation of low penetrating power such as alpha particles, the outside skin can receive some exposure without serious hazard, but for radiation that penetrates tissue readily such as X-rays, gamma rays, and neutrons, the critical parts of the body are bone marrow as blood-forming tissue, the reproductive organs, and the lenses of the eyes. The thyroid gland is important because of its affinity for the fission product iodine, while the gastro-intestinal tract and lungs are sensitive to radiation from radioactive substances that enter the body through eating or breathing.

18.2   RADIATION DOSAGE UNITS

A number of specialized terms need to be defined in order to be able to discuss the biological effects of radiation. The first is the *dose* (*D*) which is the amount of energy imparted to each gram of exposed biological tissue, and which appears as excitation or ionization of the molecules and atoms of the tissue. To illustrate, suppose that an organ weighing 50 g was exposed to radiation from a radioactive material and there resulted in a release of 0.01 J of energy. In conventional terms, the dose would be $0.01/50 = 0.0002$ J/g. A special unit that is convenient in dealing with energy absorption is the *rad*, which is 10^{-5} J/g. For the example, the dose to the organ would thus be 20 rads.

The biological effect of energy deposition may be large or small depending on the type of radiation. For instance, a rad of dosage due to fast neutrons or alpha particles is much more damaging than a rad of dosage by X-rays or gamma rays. In general, heavy particles create a more serious effect than do photons because of the greater energy loss with distance and resulting higher concentration of ionization. The "dose equivalent" (DE) is the product of the dose and a number that expresses the relative biological importance of the radiation. One of these is called a quality factor (QF) (see Table 18.1). The unit of measure of DE is the *rem* (an acronym for roentgen†-equivalent-man). For example, if QF were 3 for the radiation in the above example, the DE would be 60 rem. The millirem (1 mrem = 1/1000 rem) is a frequently used unit in describing small radiation doses.

Table 18.1.   Quality Factors.

X- and gamma rays	1
Beta particles <30 keV	1
Beta particles >30 keV	1.7
Thermal neutrons	3
Fast neutrons, protons, alpha particles	10
Heavy ions	20

The long-term effect of radiation on an organism also depends on the rate at which energy is deposited. Thus the *dose rate*, expressed in convenient units such as rads per hour or millirems per year, is used. Note that if dose is an energy, the dose rate is a power.

†The roentgen is a unit of exposure dose that was devised at a time when the principal radiations were X-rays and gamma rays. The rad and the rem are preferred units.

We shall describe the methods of calculating dosage in the following chapter. For perspective, however, we can cite some typical figures. A single sudden exposure that gives the whole body of a person 20 rem will give no perceptible clinical effect, but a dose of 400 rem will probably be fatal; natural radiation background provides about 100 mrem/yr; present medical and dental practice on the average gives nearly this same amount of additional dosage through the use of X-rays for diagnosis; regulations limit the dose rate above natural background to 5 mrem/yr at the site boundary of a power reactor.

The amounts of energy that result in biological damage are remarkably small. A gamma dose of 400 rem, which is very large in terms of biological hazard, corresponds to 4×10^{-3} J/g, which would be insufficient to raise the temperature of a gram of water as much as 0.001°C. This fact shows that radiation affects the function of the cells by action on certain molecules, not by a general heating process.

18.3 ESTABLISHMENT OF LIMITS OF EXPOSURE

A typical bottle of aspirin will specify that no more than two tablets every four hours should be administered, implying that a larger or more frequent "dose" would be harmful. Such a limit is based on experience accumulated over the years with many patients. Although radiation has medical benefit only in certain treatment, the idea of the need for a *limit* is similar.

As we seek to clean up the environment by controlling emissions of waste products from industrial plants, cities, and farms, it is necessary to specify water or air concentrations of materials such as sulfur or carbon monoxide that are below the level of danger to living beings. Ideally, there would be zero contamination, but it is generally assumed that some releases are inevitable in an industrialized world. Again, limits based on knowledge of effects on living beings must be set.

For the establishment of limits on radiation exposure, agencies have been in existence for many years. Examples are the International Commission on Radiological Protection (ICRP), the National Council on Radiation Protection and Measurements (NCRP), and the Federal Radiation Council (FRC). Their general procedure is to study data on the effects of radiation and to arrive at practical limits that take account of both the risk and benefit of using nuclear equipment and processes.

There have been many studies of the effect of radiation on animals other than man, starting with early observations of genetic effects on fruit flies. Small mammals such as mice provide a great deal of data rapidly. Since controlled experiments on man are unacceptable, most of the available information on somatic effects comes from improper practices or accidents. Data are available, for example, on the incidence of sickness and death from exposure of workers who painted radium on luminous-dial watches or of doctors who used X-rays without proper precautions. The number of serious radiation exposures in the nuclear industry is too small to be of use on a statistical basis. The principal source of information is the comprehensive study of the victims of the atomic bomb explosions in Japan in 1945. The incidence of fatalities as a function of dose is plotted on a graph similar to Fig. 18.1a, where the data are seen to lie only in the high dosage range. In the range below 10 rads, there is no statistical indication of any increase in incidence of fatalities over the number in unexposed populations. One is tempted to draw a "possible" curve as in Fig. 18.1b, taking account of the fact that no lasting effects are

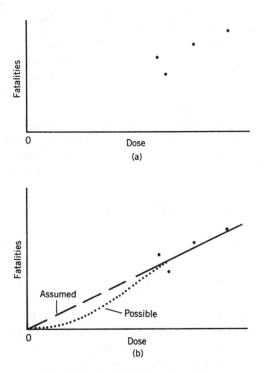

Fig. 18.1. Radiation hazard analysis.

found in many radiations. However, in order to be conservative, i.e., to overestimate effects of radiation in the interests of providing safety, a linear extrapolation through zero is made, the "assumed" curve.

There is evidence that the biological effect of a given dose administered almost instantly is greater than if it were given over a long period of time. In other words, the hazard is less for low dose rates, presumably because the organism has the ability to recover or adjust to the radiation effects. If, for example, (see Problem 18.2) the effect actually varied as the square of the dose, the linear curve would overestimate the effect by a factor of 100 in the vicinity of 1 rem. Although the hazard for low dose rates is small, and there is no clinical evidence of permanent injury, it is *not* assumed that there is a threshold dose, i.e., one below which no biological damage occurs. Instead, it is assumed that there is always some risk. The linear hypothesis is retained, in spite of the likelihood that it is overly conservative. There is a growing body of information on genetic effects in animals that tend to support this view.

The basic question then faced by standards-setting bodies is "what is the maximum acceptable upper limit for exposure?" One answer is zero, on the grounds that any radiation is deleterious. The view is taken that it is unwarranted to demand zero, as both maximum and minimum, because of the benefit from the use of radiation or from devices that have potential radiation as a byproduct.

The dose limits adopted for total body dose are 5000 mrem/yr for occupational workers, 500 mrem/yr to any individual member of the public, and an average over the whole population of 170 mrem/yr from all artificial sources other than medical applications. There are variations in permissible dose rate for workers according to the organ affected, as listed below.

Gonads, total body, and red bone marrow	5 rem/yr
Skin and bone	30 rem/yr
Other internal organs	15 rem/yr

The standards groups recommend that these figures be reduced by a factor of 10 for exposure of a person in the general public.

In the practical application of a dose rate limit standard such as the 500 and 170 mrem/yr, the question arises "what should be the limit set at the boundary of a nuclear plant?" It is clear that if 500 mrem/yr individual figure is adopted, there would be a much lower level on the average throughout the country, since nuclear plants are widely distributed geographically, and exposure would certainly decrease greatly with distance from the site. The United States Atomic Energy Commission has specified a considerably lower limit of 5 mrem/yr for the maximum dose rate at the site boundary of a nuclear power plant. Several comparisons can be made between this figure and other information. It is 1% of the individual limit or 3% of the whole population limit as recommended by ICRP and FRC. It is also about $\frac{1}{20}$ of the typical natural background. Since such low levels are not easily measured, it is necessary to calculate the dosage increase from the amount of radioactive material released. It is also comparable to the increase in dose received from cosmic radiation by a passenger on a single round trip jet airplane flight across the United States. The lower AEC standards are generally regarded as adequate for routine releases from nuclear power plants, even though zero release would be ideal. Estimates made by highly respected health physicists lead to the conclusion that the limits set by the AEC result in an exposure to the total population of the United States that statistically could result in about thirty additional deaths annually, in contrast to the millions of deaths annually due to the total of heart disease, cancer, stroke, and accidents. The effects of the slight extra exposure are believed to be completely masked by other hazards of existence.

18.4   SUMMARY

When radiation interacts with biological tissue, energy is deposited and ionization takes place that causes damage to cells. The effect on organisms is somatic, related to body health, and genetic, related to inherited characteristics. Radiation dose equivalent as a biologically effective energy deposition per gram is usually expressed in rem, with natural background giving about 0.1 rem/yr. Exposure limits are set by use of data on radiation effects at high dosages with a conservative linear hypothesis applied to predict effects at low dose rates.

18.5   PROBLEMS

18.1. A beam of 2-MeV alpha particles with current density $10^6 \, cm^{-2}$-sec, is stopped in a distance of 1 cm in air, number density $2.7 \times 10^{19} \, cm^{-3}$. How many ion pairs are formed? What fraction of the targets experience ionization?

18.2. If the chance of fatality from radiation dose is taken as 0.5 for 400 rem, by what factor would the chance at 2 rem be overestimated if the effect varied as the square of the dose rather than linearly?

18.3. A worker in a nuclear laboratory receives a whole-body exposure for

5 minutes by a thermal neutron beam at a rate 20 millirads per hour. What dose (in mrad) and dose equivalent (in mrem) does he receive? What fraction of the yearly dose limit of 500 mrem/yr for an individual is this?

18.4. A person receives the following exposures in millirems in a year: 1 medical X-ray, 100; drinking water 50; cosmic rays 30; radiation from house 60; K-40 and other isotopes 25; airplane flights 10. Find the percentage increase in exposure that would be experienced if he also lived at a reactor site boundary, assuming that the maximum AEC radiation level existed there.

19

Radiation Protection and Control

Protection of biological entities from hazard of radiation exposure is a fundamental requirement in the application of nuclear energy. Safety is provided by the use of one or more general methods that involve control of the source of radiation or its ability to affect living organisms. We shall identify these methods and describe the role of calculations in the field of radiation protection.

19.1 PROTECTIVE MEASURES

Radiation and radioactive materials are the link between a device or process as a source, and the living being to be protected from hazard. We can try to eliminate the source, or remove the individual, or insert some barrier between the two. Several means are thus available to help assure safety.

The first is to avoid the generation of radiation or isotopes that emit radiation. For example, the production of undesirable emitters from reactor operation can be minimized by the control of impurities in materials of construction and in the cooling agent. The second is to be sure that any radioactive substances are kept within containers or multiple barriers to prevent dispersal. Isotope sources and waste products are frequently sealed within one or more independent layers of metal or other impermeable substance, while nuclear reactors and chemical processing equipment are housed within leak-tight buildings. The third is to provide layers of shielding material between the source of radiation and the individual. The fourth is to restrict access to the region where the

177

radiation level is hazardous, and take advantage of the reduction of intensity with distance. The fifth is to dilute a radioactive substance with very large volumes of air or water on release, to lower the concentration of harmful material. The sixth is to limit the time that a person remains within a radiation zone, to reduce the dose received. We thus see that radioactive materials may be treated in two quite different ways— *retention* and *dispersal*, while exposure to radiation can be avoided by methods involving *distance, shielding,* and *time.*

The analysis of radiation hazard and protection and the establishment of safe practices is part of the function of the science of radiological protection or health physics. Every user of radiation must follow accepted procedures, while health physicists provide specialized technical advice and monitor the user's methods. In the planning of research involving radiation or in the design and operation of a process, calculations must be made that relate the radiation source to the biological entity, using exposure limits provided by regulatory bodies. Included in the evaluation are the necessary protective measures for known sources, or limits that must be imposed on the radiation source, the rate of release of radioactive substances, or the concentration of radioisotopes in air, water, and other materials. Although the detailed calculations are very involved, a few simple examples will help us appreciate the approach used.

19.2 ENERGY DEPOSITION

The radiation dose or dose rate to tissue of a biological specimen is of central importance. We can find the rate of energy deposition in biological tissue by use of principles discussed in Chapter 5. Visualize a stream of radiation such as gamma rays as it passes through a substance. Flux and current are the same for this stream, both j and ϕ being the product nv. The reaction rate per unit volume is $\phi\Sigma$, where the cross section is formed from components that involve absorption of the photons. If the photon energy is E, the dose rate is thus $\phi\Sigma E$, and for exposure for a time t, the dose is

$$D = \phi\Sigma Et.$$

This relation may be used to find dose, flux, or time.

For example, let us find the gamma ray flux that yields a limiting external dose of 170 mrem in 1 yr, with continuous exposure assumed. Suppose that the gamma rays have 1 MeV energy, and that the cross section for interaction with tissue of density $1.0\,\text{g/cm}^3$ is $0.03\,\text{cm}^{-1}$.

Letting the quality factor be 1 for this radiation, the dose and dose equivalent are the same, 0.170 rem or 0.170 rads. The latter as dose is $D = 1.7 \times 10^{-6}$ J/cm^3, and $E = 1$ MeV $= 1.60 \times 10^{-13}$ J. Solving for flux,

$$\phi = \frac{D}{\Sigma E t} = \frac{1.7 \times 10^{-6} \text{ J/cm}^3}{(0.03 \text{ cm}^{-1})(1.60 \times 10^{-13} \text{ J})(3.16 \times 10^7 \text{ sec})}$$

or

$$\phi = 11.2 \text{ cm}^{-2}\text{-sec}.$$

This value of the gamma ray flux may be scaled up or down if another dose limit is specified. The fluxes of various particles corresponding to 170 mrem/yr are shown in Table 19.1.

Table 19.1.   Radiation Fluxes (170 mrem/yr).

Radiation Type	Flux (cm^{-2}-sec)
X- or gamma rays	11.2
Beta particles	0.25
Thermal neutrons	5.2
Fast neutrons	0.15
Alpha particles	1.2×10^{-5}

19.3 EFFECT OF DISTANCE

For protection, advantage can be taken of the fact that radiation intensities decrease with distance from the source, varying as the *inverse square of the distance*. Let us illustrate by an idealized case of a small source, regarded as a mathematical point, emitting S particles per second, the source "strength." As in Fig. 19.1, let the rate of flow through each unit of area of a sphere of radius R about the point be labeled ϕ(cm^{-2}-sec). The flow through the whole sphere surface of area $4\pi R^2$ is then $\phi 4\pi R^2$, and if there is no intervening material, it can be equated to the source strength S. Then

$$\phi = \frac{S}{4\pi R^2}.$$

This relation expresses the inverse square spreading effect. If we have a surface covered with radioactive material or an object that emits radiation throughout its volume, the flux at a point of measurement can be found by addition of elementary contributions.

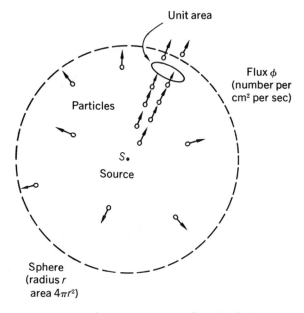

Fig. 19.1. Inverse square spreading of radiation.

Let us consider the neutron radiation at a large distance from an unreflected reactor operating at a power level of 1 MW. Since 1 W gives 3×10^{10} fissions/sec and the number of neutrons per fission is 2.5, the reactor produces 7.5×10^{16} neutrons/sec. Suppose that 20% of these escape and thus S is 1.5×10^{16} sec^{-1}. We apply the inverse square relation, neglecting attenuation in air, an assumption that would be correct for a reactor used to provide power for a spacecraft. Let us find the closest distance of safe approach to the reactor, i.e., where the neutron flux is below a safe level, say 0.15 cm^{-2}-sec, as in Table 19.1. Solving the inverse-square formula for R, we obtain

$$R = \sqrt{\frac{S}{4\pi\phi}} = \sqrt{\frac{1.5 \times 10^{16}}{(4\pi)(0.15)}} = 9 \times 10^7 \text{ cm.}$$

This is a surprisingly large distance—about 560 miles. If the same reactor were on the earth, neutron attenuation in air would reduce this figure greatly, but the necessity for shielding by solid or liquid materials is clearly revealed by this calculation.

As another example, let us find how much radiation is received at a distance of 1 mile from a nuclear power plant, if the dose rate at the plant boundary, $\frac{1}{4}$-mile radius, is 5 mrem/yr. Neglecting attenuation in air, the inverse-square reduction factor is $\frac{1}{16}$ giving 0.03 mrem/yr.

19.4 SHIELDING

The evaluation of necessary protective shielding from radiation makes use of the basic concepts and facts of radiation interaction with matter described in Chapter 5. Let us consider the particles with which we must deal. Since charged particles—electrons, alpha particles, protons, etc.— have a very short range in matter, attention needs to be given only to the penetrating radiation—gamma rays (or X-rays) and neutrons. The attenuation factor with distance of penetration for photons and neutrons may be expressed in exponential form $e^{-\Sigma r}$, where r is the distance from source to observer and Σ is an appropriate macroscopic cross section. Now Σ depends on the number of target atoms, and through the microscopic cross section σ also depends on the type of radiation, its energy, and the chemical and nuclear properties of the target.

For fast neutron shielding, a light element is preferred because of the large neutron energy loss per collision. Thus hydrogenous materials such as water, concrete, or earth are effective shields. The objective is to slow neutrons within a small distance from their origin and to allow them to be absorbed at thermal energy. Thermal neutrons are readily captured by many materials, but boron is preferred because accompanying gamma rays are very weak.

Let us compute the effect of a water shield on the fast neutrons from the example reactor used earlier. The macroscopic cross section appearing in the exponential formula $e^{-\Sigma r}$ is now called a "removal cross section," since many fast neutrons are removed from the high-energy region by one collision with hydrogen, and eventually are absorbed as thermal neutrons. Its value for fission neutrons in water is around 0.10 cm^{-1}. A shield of thickness 8 ft = 244 cm would provide an attenuation factor of $e^{-24.4} = 10^{-10.6} = 2.5 \times 10^{-11}$. The inverse-square reduction with distance is

$$\frac{1}{4\pi r^2} = \frac{1}{4\pi(244)^2} = 1.3 \times 10^{-6}.$$

The combined reduction factor is 3.2×10^{-17}; and with a source of 1.5×10^{16} neutrons/sec, the flux is down to 0.5 neutrons/cm^2-sec, which is only slightly higher than the safe level. The addition of an extra foot of water shield would provide adequate protection, for steady reactor operation at least.

For gamma ray shielding, in which the main interaction takes place with atomic electrons, a substance of high atomic number is desired. Compton scattering varies as Z, pair production as Z^2, and the photoelectric effect as Z^5. Elements such as iron and lead are particularly useful for gamma shielding. The amount of attenuation depends on the material of the shield, its thickness, and the photon energy. The literature gives values of the "mass attenuation coefficient," which is the quotient of the macroscopic cross section and the material density. Typical values for a few elements at different energies are shown in Table 19.2. For 1 MeV gamma rays in iron, density 7.8 g/cm³, we deduce that Σ is 0.467 cm⁻¹. For water, with $\frac{1}{9}$ g of hydrogen and $\frac{8}{9}$ g of oxygen per cubic centimeter, the mass attenuation coefficient is $(1/9)(0.126) + (8/9)(0.0637) = 0.0706$ cm²/g; with density 1.0 g/cm³, Σ is 0.0706 cm⁻¹. To achieve the same reduction in a beam of gammas, the thickness of an iron shield can be about $\frac{1}{6}$ that of a water shield.

As an example of gamma shielding calculations, we estimate the thickness of lead shield that should be provided for a point source of strength 1 millicurie (3.7×10^7 d/sec), emitting 1 MeV gamma rays, in order to bring the continuous exposure dose rate down to a level of 5 mrem/yr at the surface of the shield. From our previous calculations or Table 19.1, this corresponds to 0.33 cm⁻²-sec. We must take account of the fact that the simple exponential attenuation relation refers only to the transmission of gamma rays that have made no collision. Those which are scattered by the Compton effect can return to the stream and contribute to the dose, as sketched in Fig. 19.2. To account for this "buildup" of radiation, a *buildup factor B* depending on Σr is calculated. Figure 19.3 shows B for 1 MeV gammas in lead. The microscopic cross section for this radiation is found from Fig. 6.4 to be 25 barns. For lead with atomic

Table 19.2.   Mass Attenuation Coefficients (cm²/g).

Element Energy (MeV)	H	O	Fe	Pb
0.01	0.385	5.78	173	133
0.1	0.294	0.156	0.370	1.91
1	0.126	0.0637	0.0599	0.0776
10	0.0325	0.0209	0.0298	0.0506

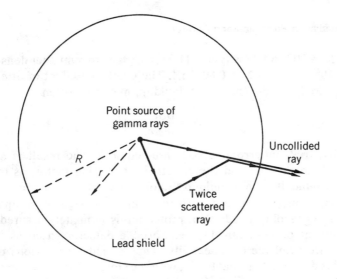

Fig. 19.2. Buildup effect.

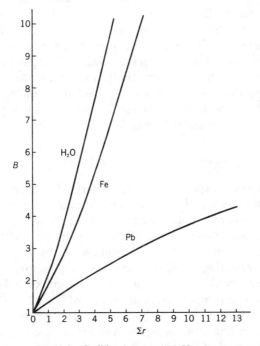

Fig. 19.3. Buildup factors, 1 MeV gammas.

weight $M = 207$ and density $\rho = 11.3$ g/cm^3, the atom number density N is 0.033×10^{24} cm^{-3} and Σ is 0.80 cm^{-1}. The combined effect of attenuation with material and distance, with buildup, may be written

$$\phi = \frac{BSe^{-\Sigma r}}{4\pi r^2}.$$

To find r, trial-and-error methods are required. The result is approximately 15 cm or 6 in., and the effective attenuation factor, as the ratio ϕ/S, is around 10^{-8}. The buildup factor turns out to be about 4.

Although calculations are performed in the design of equipment or experiments involving radiation, protection is ultimately assured by the measurement of radiation dosage. Portable detectors used as "survey meters" are available commercially. They employ the various detector principles described in Chapter 11, with the Geiger–Muller counter having the greatest general utility. Special detectors are installed to monitor general radiation levels or the amount of radioactivity in effluents.

The possibility of accidental exposure to radiation always exists in a laboratory or plant, in spite of all precautions. In order to have information immediately, personnel wear dosimeters, which are pen-size self-reading ionization chambers that detect and measure dose. For a more permanent record, film badges are worn. These consist of several photographic films of different sensitivity, with shields to select radiation types. They are developed periodically, and if significant exposure is noted, individuals are relieved of future work in areas with potential radiation hazards for a suitable length of time.

19.5 INTERNAL EXPOSURE

We now turn to the exposure of internal parts of an organism as a result of having taken in radioactive substances. Special attention will be given to the human body, but similar methods will apply to other animals and even to plants. Radioactive materials can enter the body by drinking, breathing, or eating, and to a certain extent can be absorbed through pores or wounds. The resulting dosage depends on many factors: (a) the amount that enters, which in turn depends on the rate of intake and elapsed time; (b) the chemical nature of the substance, which affects affinity with molecules of particular types of body tissue and which determines the rate of elimination (the term biological half-life is used in this connection, being the time for half of an initial amount to be removed); (c) the particle size, which relates to progress of the material through the body; (d) the radioactive half-life, the energy, and kind of radiation, which determine the activity and energy deposition rate, and the length of time the radiation exposure persists; (e) the radiosensitivity of the tissue, with the gastro-intestinal tract, reproductive organs, and bone marrow as the most important.

The maximum permissible concentration of an isotope in air or water can be calculated by consideration of all the above factors. Such an analysis in general is quite detailed, but there are some special cases in which a rough estimate can be made easily. An example is a gaseous fission product such as krypton, or the isotope tritium, half-life 12.3 yr, which is produced in a reactor in several ways. Suppose that a person is surrounded by air that contains a radioactive gas. If this "cloud" is large in comparison with radiation mean free paths or ranges, the rate of absorption of energy is equal to the rate of emission of energy. The beta particles of tritium have a low average energy, 0.006 MeV, and their range in air is very short.

Let us calculate the maximum permissible concentration (MPC) of tritium in air of density 1.29×10^{-3} g/cm^3, expressed as an activity per unit volume (μCi/cm^3), that will yield a dose rate of 5 mrem/yr. The energy released per second is

$$(\text{MPC } \mu\text{Ci/cm}^3)(3.7 \times 10^4 \text{ d/sec-}\mu\text{Ci})(0.006 \text{ MeV})(1.60 \times 10^{-13} \text{ J/MeV}),$$

while the energy absorbed per second is

$$\frac{(5 \times 10^{-3} \text{ rad/yr})(10^{-5} \text{ J/g-rad})(1.29 \times 10^{-3} \text{ g/cm}^3)}{3.16 \times 10^7 \text{ sec/yr}}.$$

Equating and solving, MPC $= 5.7 \times 10^{-8}$ μCi/cm^3. Although this corresponds to a very small number of particles in comparison with the 2.7×10^{19} cm^{-3} of air molecules, it turns out to be large in comparison with MPC values for more hazardous isotopes such as radium-226 or plutonium-239.

19.6 RADIATION DAMAGE

Protection of living beings from the deleterious effects of radiation is of primary concern in the operation of any nuclear device, but attention must also be given to the possibility of radiation damage to nonliving materials. As discussed in Chapter 6, electrons, gamma rays, neutrons, and heavy charged particles can cause excitation, ionization, and dissociation of chemical substances, thus rendering them unsuitable for their original purpose. Generally speaking, organic materials such as plastics are sensitive to beta and gamma radiation, i.e., are most readily damaged. Their chemical bonds are relatively weak and easily broken. In contrast, metals are resistant to such radiation because their conduction electrons can absorb much of the energy of radiation without experiencing a structural change. However, metals are affected by neutron bombardment, principally through atom displacements. Thus metals in a nuclear reactor exhibit increases in hardness and tensile strength, with a loss in ductility. The effects depend on the dose, proportional to the total number of fast neutrons to which the material has been exposed. If the flux is $\phi = nv$, in a time t, the exposure is nvt, called the integrated flux or fluence. In the range of fluences 10^{19}–10^{21} per square centimeter most metals show such changes. The degree of damage tends to be smaller at higher material temperatures, since displaced atoms move about more readily and return to their original sites.

The success of the liquid metal fast breeder reactor will be determined in part by the ability of the fuel rods to withstand the rigorous thermal, mechanical, and radiation environment. Under the effects of radiation, the stainless steel cladding tends to swell and undergo creep, a slow stretching process, while the oxide fuel and its fission products interact with the cladding.

19.7 SUMMARY

Radiation protection of living organisms requires control of sources, barriers between sources and organism, or removal of the biological entity. Calculations required to evaluate external hazard include the dose as it depends on radiation flux and energy, material, and time; the inverse square spreading effect; and the exponential attenuation in shielding materials. Internal hazard depends on many physical and biological factors. Maximum permissible concentrations of radioisotopes in air or water can be deduced from the properties of the emitter and the specified dose rate limits. Radiation damage in nonliving substances is of concern, since organic chemicals are readily affected, and neutron bombardment changes reactor metal properties.

19.8 PROBLEMS

19.1. What is the rate of exposure in mrem/yr corresponding to a continuous gamma ray flux of 100 cm^{-2}-sec? What dose equivalent would be received by a person who worked 40 hr/wk throughout the year in such a flux?

19.2. A Co-60 source is to be selected to test radiation detectors for operability. Assuming that the source can be kept at least 1 m from the body, what is largest strength acceptable (in μCi) to assure an exposure rate of less than 500 mrem/yr? (Note that two gammas of energy 1.1 and 1.3 MeV are emitted.)

19.3. By comparison with the tritium analysis, estimate the MPC in air for Kr-85, average beta particle energy 0.22 MeV.

19.4. The nuclear reactions resulting from thermal neutron absorption in boron and cadmium are

$$^{10}_{5}B + ^{1}_{0}n \rightarrow ^{7}_{3}Li + ^{4}_{2}He,$$

$$^{113}_{48}Cd + ^{1}_{0}n \rightarrow ^{114}_{48}Cd + \gamma(5 \text{ MeV}).$$

Which material would you select for a radiation shield? Explain.

19.5. Find the gamma ray flux that gets through a spherical lead shield of 4-in. radius if the source of 1 MeV gammas is of strength 200 mCi.

19.6. The maximum permissible concentration (MPC) of unidentified beta- and gamma-emitting isotopes in water is $10^{-7}\ \mu$Ci/cm^3. In order to assure that the actual release is no more than 1% of the MPC, a limitation on the discharge rate (r) in gallons per minute (gpm) must be applied for each radioactive solution of specific activity c(μCi/cm^3). Assuming a further dilution of any fluid released by a river stream flow of 1500 gpm, develop a working formula relating r to c, and plot a graph for convenient use. Suggestion: 3-cycle log–log paper.

20
Reactor Safety

It is well known that the accumulated fission products in a reactor that has been operating for some time constitute a potential source of radiation hazard. Assurance is needed that the integrity of the fuel is maintained throughout the operating cycle, with negligible release of radioactive materials. This implies limitations on power level and temperature, and adequacy of cooling under all conditions. Fortunately, inherent safety is provided by physical features of the fission chain reaction. In addition, the choice of materials, their arrangement, and restrictions on modes of operation give a second level of protection. Devices and structures that minimize the chance of accident and the extent of radiation release in the event of accident are a third line of defense. Finally, nuclear plant location at a distance from centers of high population density results in further protection.

We shall now describe the dependence of numbers of neutrons and reactor power on the multiplication factor, which is in turn affected by temperature and control rod absorbers. Then we shall examine the precautions taken to prevent release of radioactive materials to the surroundings and discuss the philosophy of safety.

20.1 NEUTRON POPULATION GROWTH

The multiplication of neutrons in a reactor can be described by the effective multiplication factor k, as discussed in Chapter 12. The introduction of 1 neutron produces k neutrons, they in turn produce k^2, and so on. Such a behavior would seem to be completely analogous to the increase in principal with compound interest or the exponential growth of human population. We shall see shortly that there are some important differences, but it will be desirable to develop an expression for the growth with time for *any* population n, if for each individual the gain per cycle of reproduction is $\delta k = k - 1$ and the time for 1 cycle is l. In an infinitesimal time dt, the gain with n individuals is dn, and by proportions

$$\frac{dn}{dt} = \frac{\delta k n}{l}.$$

If the coefficient of n on the right side is constant, and if the number present at time zero is n_0, the solution is found to be

$$n = n_0 e^{t/T},$$

where T is the "period," the time for the population to increase by a factor $e = 2.718\ldots$, given by $T = l/\delta k$. When applied to people, the formula states that the population grows more rapidly the more frequently reproduction occurs and the more abundant the progeny.

A typical cycle time l for neutrons in a thermal reactor is very short, around 10^{-5} sec, so that a δk as small as 0.02 would give a very short period of 0.0005 sec. The growth according to the formula would be exceedingly rapid, and if sustained would consume all of the atoms of fuel in a fraction of a second.

A peculiar and fortunate fact of nature provides an inherent reactor control for values of δk in the range 0 to around 0.0065. Recall that around 2.5 neutrons are released from fission. Of these, some 0.65% appear later as the result of radioactive decay of a certain group of fission products, and are thus called *delayed neutrons*. The average half-life of the isotopes from which they come, taking account of their yields, is around 8.8 sec. This corresponds to a mean life $\tau = t_H/0.693 = 12.7$ sec., as the average length of time required for a radioactive isotope to decay. Although there are very few delayed neutrons, their presence extends the cycle time greatly and slows the rate of growth of neutron population. To understand this effect, let β be the fraction of all neutrons that are delayed, a value 0.0065 for U-235; $1 - \beta$ is the fraction of those emitted instantly as "prompt neutrons." If the length of time before the delayed neutrons appear is τ, but the prompt neutrons appear instantly, the average delay is $\beta\tau + (1 - \beta)\,0 = \beta\tau$. Now since $\beta = 0.0065$ and $\tau = 12.7$ sec, the product is 0.083 sec, greatly exceeding the multiplication cycle time, which is only 10^{-5} sec. The delay time can thus be regarded as the effective generation time, $\bar{l} = \beta\tau$. This approximation holds for values of δk much less than β.

For example, let $\delta k = 0.001$, and use $\bar{l} = 0.083$ sec in the exponential formula. In one second $n/n_0 = e^{0.012} = 1.01$, a very slight increase.

On the other hand, if δk is much larger than β we still find very rapid responses, even with delayed neutrons. If all neutrons were prompt, 1 neutron would give a gain of δk, but since the delayed neutrons actually appear much later, they cannot contribute to the immediate response. The apparent δk is then $\delta k - \beta$, and the cycle time is l. We can summarize by listing the period T for the two regions.

$$\delta k \ll \beta, \qquad T \simeq \frac{\beta \tau}{\delta k},$$

$$\delta k \gg \beta, \qquad T \simeq \frac{l}{\delta k - \beta}.$$

Even though β is a small number, it is conventional to consider δk small only if it is less than 0.0065 but large if it is greater. Figure 20.1 shows the

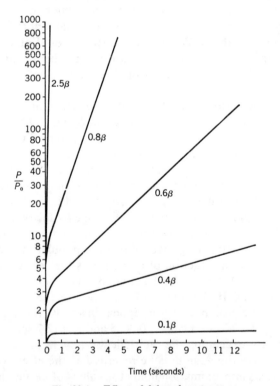

Fig. 20.1.   Effect of delayed neutrons.

growth in reactor power for several different values of reactivity ρ, defined as $\delta k/k$. Since k is close to 1, $\rho \simeq \delta k$. We conclude that the rate of growth of neutron population or reactor power is very much smaller than expected, so long as δk is kept well below the value β, but that rapid growth will take place if δk is larger than β.

We have used the value of β for U-235 for illustration, but should note that its effective value depends on reactor size and type of fuel, e.g., β for Pu-239 is only 0.0021. Also, the value of neutron cycle time depends on the energy of the predominant neutrons. The l for a fast reactor is much shorter than that for a thermal reactor.

20.2 TEMPERATURE EFFECTS

Reactor safety is also enhanced by effects involving changes in temperature of moderator and fuel. As designed, reactors have inherent safety against excessive power increases because of a self-regulation effect. An increase in power produced by an applied reactivity tends to cause a temperature increase of the materials, which gives rise to a negative reactivity, canceling out the initial value (see Fig. 20.2).

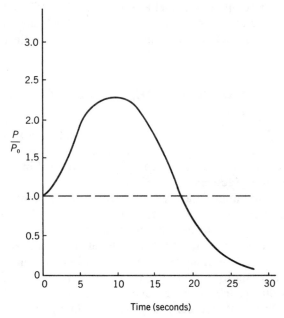

Fig. 20.2. Effect of temperature on power.

Depending on the reactor type, advantage is taken of several physical effects. In a small research-training reactor with light water moderator, the thermal expansion of water allows greater neutron leakage and the multiplication factor is reduced. In a large power reactor where neutron leakage is relatively small, the effect of temperature on resonance absorption is exploited. A temperature rise causes fuel atoms to have higher speeds, and the rate of interaction of moving nuclei and neutrons is changed. This Doppler effect, so named in analogy to frequency changes in sound or light when there is relative motion of the source and observer, tends to increase the amount of resonance absorption. The factor p in the multiplication is reduced.

The net reactivity feedback accounting for all effects can be expressed in terms of the *temperature coefficient of reactivity*, α, or the *power coefficient*, a, by writing

$$\rho = \alpha \Delta T$$

or

$$\rho = a \Delta P,$$

where the constants of proportionality are respectively the reactivity resulting from a unit change in temperature or power. For example, suppose that the temperature coefficient were $-10^{-4}/°C$. A temperature rise of 50°C would give a reactivity of -0.005, which is almost as large in magnitude as β, and could thus compensate for a rather large reactivity addition.

20.3 SAFETY THROUGH DESIGN, FABRICATION, AND OPERATION

Even though a reactor is relatively insensitive to increases in multiplication in the region below β and temperature increases tend to shut the reactor down, additional protection is provided in the design and through operation practices.

Most reactors contain an array of movable rods of neutron-absorbing material. Figure 20.3 shows the arrangement for a PWR type. The rods serve four purposes: (a) to permit temporary increases of multiplication that bring the reactor up to desired power level or to make adjustments to the power; (b) to cause changes in the flux and power shape within the core, usually striving for uniformity; (c) to compensate for effects of fuel consumption during the operating cycle of the reactor; and (d) to shut down the reactor automatically or manually in the event of unusual

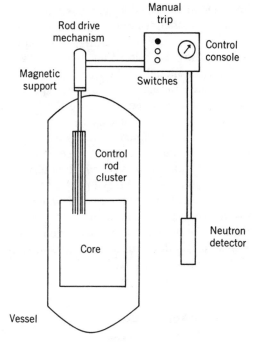

Fig. 20.3.   Reactor control.

behavior. To ensure the effectiveness of the shutdown role, a certain
group of "safety rods" are kept withdrawn from the reactor at all times
during operation. In the PWR, they are supported by electromagnets that
release the rod on interruption of current while in the BWR they are
driven in from the bottom of the vessel by hydraulic means.

Since almost all of the radioactivity generated by a reactor appears in
the fuel elements, great precautions are taken to assure the integrity of the
fuel. Care is taken in fuel fabrication plants to produce fuel pellets that
are identical chemically, of the same size and shape, and of common
U-235 concentration. If one or more pellets of unusually high fissile
material content were used in a reactor, excessive local power production
and temperature would result. The metal tubes that contain the fuel
pellets are made sufficiently thick to stop the fission fragments, to provide
the necessary mechanical strength to support the column of pellets, and to
withstand erosion by water flow or corrosion of water at high
temperatures. Also, the tube must sustain a variable pressure difference

caused by moderator-coolant outside and fission product gases inside. The "cladding" material usually selected for low neutron absorption and for resistance to chemical action, melting, and radiation damage in thermal reactors is zircaloy, an alloy that is about 98% zirconium with small amounts of tin, iron, nickel, and chromium. The tube is formed by an extrusion process that eliminates seams, and special fabrication and inspection techniques are employed to assure that there are no defects such as deposits, scratches, holes, or cracks.

Each reactor has a set of specified limits on operating parameters to assure protection against events that could cause hazard. Typical of these is the upper limit on total reactor power, which determines temperatures throughout the core. Another is the ratio of peak power to average power, which is related to hot spots and to cladding integrity. A third is the minimum coolant flow rate. Maintenance of chemical purity of the coolant to minimize corrosion, limitation on allowed leakage rate from the primary cooling system, and continual observations on the level of radioactivity in the coolant serve as further precautions against release of radioactive materials.

20.4 STANDARDS, QUALITY ASSURANCE, LICENSING, AND REGULATION

In the foregoing paragraphs we have alluded to a few of the physical features and procedures employed in the interests of safety. These have evolved from experience over some twenty years, as described in Chapter 17, and much of the design and operating experience has been translated into widely used *standards*, which are descriptions of acceptable practice. Professional technical societies, industrial organizations, and the federal government cooperate in the development of these useful documents.

In addition, requirements related to safety have a legal status, since all safety aspects of nuclear systems are rigorously regulated by federal law, administered by the United States Atomic Energy Commission (AEC). Before a prospective owner of a nuclear plant can receive a permit to start construction, he must submit a comprehensive preliminary safety analysis report (PSAR) and an environmental impact statement. Upon approval of these, a final safety analysis report (FSAR), technical specifications, and operating procedures must be developed in parallel with the manufacture and construction. An exhaustive testing program of components and systems is carried out at the plant. The documents and test results form the basis for an operating license.

Throughout the analysis, design, fabrication, construction, testing and operation of a nuclear facility, adequate *quality control* (QC) is required. This consists of the careful documented inspection of all steps in the sequence. In addition, a *quality assurance* (QA) program that verifies that quality control is being exercised properly is imposed. Licensing by the AEC is possible only if the QA program has satisfactorily performed its function. During the life of the plant, periodic inspections of the operation are made by the AEC to ascertain whether or not the owner is in compliance with safety regulations, which include special and regular reports from the operating organization to the AEC.

20.5 EMERGENCY CORE COOLING

The design features and operating procedures for a reactor are such that under normal conditions a negligible amount of radioactivity will get into the coolant and find its way out of the primary loop. Knowing that abnormal conditions can exist, the worst possible event, called a design basis accident, is postulated. Backup protection equipment, called engineered safety features, is provided to render its effect negligible. A loss of coolant accident (LOCA) is the condition typically assumed, in which the main coolant piping somehow breaks and thus the pumps cannot circulate coolant through the core. Although the reactor power would be reduced immediately by use of safety rods in such a situation, there is a continuing supply of heat from the decaying fission products that would tend to increase temperatures above the melting point of the fuel and cladding. In a severe situation, the fuel tubes would be damaged, and a considerable amount of fission products released. In order to prevent melting, an emergency core cooling system (ECCS) is provided in water-moderated reactors, consisting of auxiliary pumps that inject and circulate cooling water to keep temperatures down. The operation of a typical ECCS can be understood by study of some schematic diagrams.

The basic reactor system (Fig. 20.4) includes the reactor vessel, the primary coolant pump, and the steam generator, all located within the containment building. The system actually may have more than one steam generator and pump—these are not shown for ease in visualization. We show in Fig. 20.5 the auxiliary equipment that constitutes the engineered safety (ES) system. First is the *high-pressure injection system,* which goes into operation if the vessel pressure expressed in pounds per square inch (psi) drops from a normal value of around 2250 psi to about 1500 psi as the result of a small leak. Water is taken from the borated water storage tank

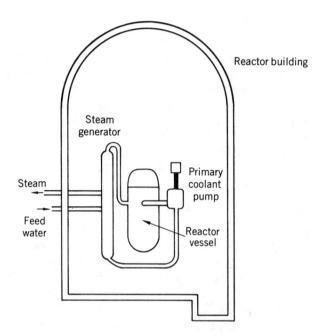

Fig. 20.4.   Reactor containment.

and introduced to the reactor through the inlet cooling line. Next is the
core flooding tank, which delivers borated water to the reactor through
separate nozzles in the event a large pipe break occurs. Such a rupture
would cause a reduction in vessel pressure and an increase in building
pressure. When these become around 600 and 4 psi, respectively, the
water enters the core through nitrogen pressure in the tank. Then if the
primary loop pressure falls to around 500 psi, the *low-pressure injection*
pumps start to transfer water from the borated water storage tank to the
reactor. When this tank is nearly empty, the pumps take spilled water
from the building sump as a reservoir and continue the flow, through
coolers that remove the decay heat from fission products. Another device,
the building spray system, also goes into operation if the building pressure
increases. It takes water from the sump and discharges it from a set of
nozzles located above the reactor, in order to provide a means for
condensing steam and certain released fission products. At the same time,
the reactor building emergency cooling units are operated to reduce the
temperature of any released vapor.

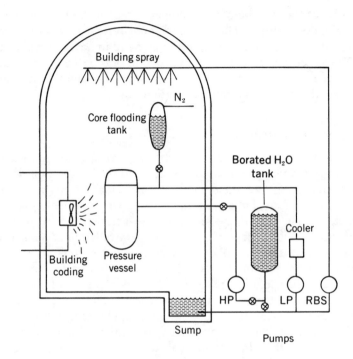

Fig. 20.5.   Emergency core cooling system.

We can estimate the magnitude of the problem of removing fission product heat. For a reactor fueled with U-235, operated for a long time at power P_0 and then shut down, the power associated with the decay of accumulated fission products is $P_f(t)$, given by an empirical formula such as

$$P_f(t) = P_0 A t^{-a}.$$

For times larger than 10 sec after reactor shutdown the decay is represented approximately by using $A = 0.066$ and $a = 0.2$. We find that at 10 sec the fission power is 4.2% of the reactor power. By the end of a day, it has dropped to 0.68%, which still corresponds to a sizeable power, viz., 20 MW for a 3000 MWt reactor. The ECCS must be capable of limiting the surface temperature of the zircaloy cladding to specified values, e.g., 2200°F, of preventing significant chemical reaction, and of maintaining cooling over the long term after the postulated accident.

The role of the steel-reinforced concrete reactor building is to provide containment of fission products that might be released from the reactor. It is designed to withstand internal pressures and to have a very small leak rate. The reactor building is located within a zone called an exclusion area, of radius of the order of a quarter of a mile, and the nuclear plant site is several miles from any population center.

20.6 THE NUCLEAR CONTROVERSY

The use of nuclear energy for electrical power production has been a subject of controversy in a number of areas of the United States. There is complete agreement that the radiation from radioactive fission products is extremely hazardous, but a great range of views as to the implications of that fact. At one end of the spectrum are activists who reject nuclear energy outright and seek to ban or delay the installation of nuclear plants. Next are a very small number of scientists and writers who choose to interpret data to suit their preconceived opinions as to the present or future effects of radiation release and stress the lack of knowledge or administrative control of nuclear processes. Third are informed but concerned scientists, engineers, and citizens who feel that the consequences of a major nuclear accident are so great that additional controls are needed or that other energy sources should be developed. Fourth are those technical and administrative people in government and industry who are charged with assuring continuous, adequate, inexpensive power for the public, and are satisfied that the very elaborate system of regulation and physical devices is adequate. The principal questions that prompt disagreement are whether or not reactors are safe and whether or not their benefit exceeds the risk.

20.7 PHILOSOPHY OF SAFETY

The subject of safety is a subtle combination of technical and psychological factors. Regardless of the precautions that are provided in the design, construction, and operation of any device or process, the question can be raised "Is it safe?". The answer cannot be a categorical "yes" or "no," but must be expressed in more ambiguous terms related to the chance of malfunction or accident, the nature of protective systems, and the consequences of failure. This leads to more philosophical questions such as "How safe is safe?" and "How safe do we want to be?".

Every human endeavor is accompanied by a certain risk of loss or damage or hazard to individuals. In the act of driving an automobile on the highways, or in turning on an electrical appliance in the home, or even in the process of taking a bath, one is subject to a certain danger. Everybody agrees that the consumer deserves protection against hazard outside his personal control, but it is not at all clear as to what lengths it is necessary to go. In the absurd limit, for instance, a complete ban on all mechanical conveyances would assure that no one would be killed in accidents involving cars, trains, airplanes, boats, or spacecraft. Few would accept the restrictions thus implied. It is easy to say that reasonable protection should be provided, but the word "reasonable" has different meanings among people. The concept that the benefit must outweigh the risk is appealing, except that it is very difficult to assess the risk of an innovation for which no experience or statistical data are available, or for which the number of accidents is so low that many years would be required for adequate statistics to be accumulated. Nor can the benefit be clearly defined. A classic example is the use of a pesticide that assures protection of the food supply for many, with finite danger to certain sensitive individuals. To the person affected adversely, the risk completely overshadows the benefit. The addition of safety measures is inevitably accompanied by increased cost of the device or product, and the ability or willingness to pay for the increased protection varies widely among people.

It is thus clear that the subject of safety falls within the scope of the social-economic-political structure and processes and is intimately related to the fundamental conflict of individual freedoms and public protection by control measures. It is presumptuous to demand that every action possible should be taken to provide safety, just as it is negligent to contend that because of evident utility, no effort to improve safety is required. Between these extreme views, there remains an opportunity to arrive at satisfactory solutions, applying technical skill accompanied by responsibility to assess consequences. It is most important to provide understandable information, on which the public and its representatives can base judgments and make wise decisions as to the proper level of investment of effort and funds.

20.8 SUMMARY

Prevention of release of radioactive fission products and fuel isotopes is the ultimate purpose of safety features. Inherent reactor safety is provided by delayed neutron and temperature effects. Control rods permit quick shutdown, and reactor components are designed and constructed to minimize the chance of failure, with licensing administered by a federal agency. Equipment is installed to reduce the hazard in the event of a postulated accident. Controversy about nuclear energy is related to safety and the relation of benefit and risk.

20.9 PROBLEMS

20.1. (a) If the total number of neutrons from fission by thermal neutrons absorbed in U-235 is 2.43, how many are delayed and how many are prompt?
(b) A reactor is said to be "prompt critical" if it has a positive reactivity of β or more. Explain the meaning of the phrase.
(c) What is the period for a reactor with neutron cycle time 5×10^{-6} sec if the reactivity is 0.013?
(d) What is the period if instead the reactivity is 0.0013?

20.2. A reactor is operating at a power level of 250 MWe. Control rods are removed to give a reactivity of 0.0005. Noting that this is much less than β, calculate the time required to go to a power of 300 MWe, neglecting any temperature feedback.

20.3. When a large positive reactivity is added to a fast reactor assembly, the power rises to a peak value and then drops, crossing the initial power level. In this response, which is the result of a negative temperature effect, the times required for the rise and fall are about the same. If the neutron cycle time is 4×10^{-6} sec, what would be the approximate duration of an energy pulse resulting from a reactivity of 0.0165, if the peak power is 10^3 times the initial power? See figure.

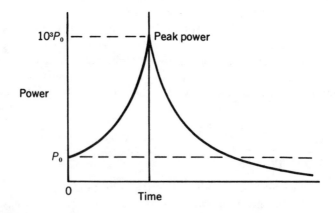

20.4. During a "critical experiment," in which fuel is initially loaded into a reactor, a fuel element of reactivity worth 0.002 is suddenly dropped into a core that is already critical. If the temperature coefficient is $-5 \times 10^{-5}/°F$, how high will the temperature of the system go above room temperature before the positive reactivity is canceled out?

20.5. How long will it take for a fully withdrawn control rod in a reactor of height 12 ft to drop into a reactor core neglecting all friction and buoyancy effects? (Recall $s = \frac{1}{2}gt^2$ with $g = 32.2$ ft/sec^2.)

20.6. Calculate the ratio of fission product power to reactor power for four times after shutdown—1 day, 1 week, 1 month, and 1 year, using the approximation $A = 0.066$, $a = 0.2$.

21

Radioactive Waste Processing and Disposal

The byproducts of every neutron-induced fission of a heavy element are two lighter fragments that are usually highly radioactive, with half-lives ranging from fractions of a second to thousands of years. There being no practical way in sight to render these fission products nonradioactive and therefore inert, we face the fact that the use of nuclear energy is accompanied by a continuing demand for safe handling, transportation, processing, storage, and disposal of potentially hazardous materials. Each of these steps will be reviewed in this chapter.

21.1 AMOUNTS OF FISSION PRODUCTS

An appreciation can be gained of the magnitude of the problem of handling reactor-produced radioactive materials by a study of their physical characteristics. First, we note that the weight and volume of fission products are rather small. When a U-236 nucleus splits, the mass-energy released is only $200 \text{ MeV}/(931 \text{ MeV/amu}) \simeq 0.2$ amu. The atomic masses of fission products still add up to approximately 236, and thus we can essentially equate the weights of fuel fissioned and waste products. For each megawatt-day of energy release, 1.3 g of U-235 are consumed. Of this, 85% is fission, so the amount of U-235 fissioned is $(0.85)(1.3) = 1.1$ g/MWd.

A reactor operating at 3000 MWt thus produces 3.3 kg or 7.3 lb of fission products per day or about 2650 lb/yr. If we assume an average specific gravity of 10, i.e., 624 lb/ft³, the volume comes out 4.2 ft³/yr. The amount of material to be handled initially is much larger, of course, being the total including unused fuel and tubing. On the other hand, the amount of pure radioactive material is much lower than this figure at any time, because of prior decay of short-lived isotopes.

Although the actual amount of material is small, the heat generation rate, the activity, and the resultant radiation level is high for a considerable time after fuel is removed from the reactor. We may again apply the decay heat formula (Chapter 20) to estimate the requirements on cooling and shielding. The power from decay at a time as long as 3 months (7.8×10^6 sec) after shutdown of a 3000-MW reactor is $P = (3000)(0.066)(7.8 \times 10^6)^{-0.2} = 8.3$ MW. If we assumed that typical particles released have an energy of 1 MeV, this would correspond to 1.4×10^9 Ci.

21.2 FUEL HANDLING, TRANSPORTATION, AND PROCESSING

The sequence of steps required in the processing of "spent" fuel and an indication of the precautions taken is now given. The fuel cycle for typical light water power reactors involves the removal of one-third of the irradiated fuel each year, the insertion of one-third fresh fuel, with rearrangement to optimize power production. Thus we start with fuel elements that have been irradiated for three years, during which time much of the U-235 has been burned, Pu-239 has been generated and partially consumed, and a great variety of chemical species that constitute the fission products has been deposited. At the end of its useful life, the spent fuel is removed from the reactor and transferred to storage. It is kept under water, which provides the necessary protective shielding from gamma rays and permits residual heat from decay to be dissipated. Over a period of several months of storage, the radiation level declines to values safe for transportation of the fuel elements to a fuel reprocessing plant. Figure 21.1 shows the sequence of events.

The shipment of fuel requires protection against the release of radioactivity, exposure to external radiation, and accidental criticality. The shipping containers, or fuel casks, consist of steel tanks weighing up to 50 tons, about 18 ft long and 5 ft in diameter. Internal lead-lined steel tubes serve to isolate fuel assemblies from one another; the vessel contains water to provide cooling and shielding, and it is sealed to prevent escape of radioactive materials; fins on the outside of the cask help remove heat during shipment.

The casks are designed to meet certain tests to assure that an accident

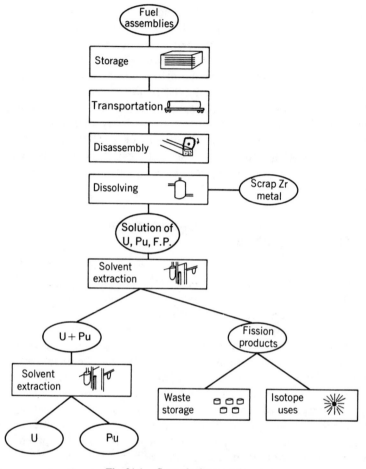

Fig. 21.1.   Spent fuel processing.

during transit by rail or truck does not result in contamination. The casks are constructed in such a way that they will withstand the rigors of a free fall of 30 ft to a hard flat surface or a 40-in. drop onto a 6-in. diameter steel pin, or exposure to a gasoline fire at temperature 1475°F for half an hour, followed by immersion in water for 8 hr.

On arrival at the chemical plant where fuel reprocessing is done (Fig. 21.1), the fuel elements are chopped up by mechanical shears, and the fuel pellets containing uranium, plutonium, and fission products are dissolved by nitric acid. The zircaloy cladding is set aside, and the solution treated by a solvent-extraction process to separate out the uranium and plutonium. The uranium is converted from the nitrate to the oxide and then to the hexafluoride for shipment to the isotope separation plant. The plutonium is put in concentrated solution for return to storage or to a fuel fabrication center. Special precautions in handling plutonium are required for three reasons: (a) it is a valuable fissile material, superior to U-235 in some respects; (b) its half-life of 24,390 yr and its affinity for bone tissue make it a particularly hazardous material; and (c) the element is in a form suitable for use in nuclear weapons and thus safeguards must be imposed to prevent unauthorized diversion to unfriendly foreign governments or to groups with illegal intent. Elaborate technical and administrative controls are applied to provide an accurate knowledge of material inventory or possible diversion at every stage of handling. In addition to chemical sampling, measurements of gamma rays and delayed neutrons are made.

During the processing, special attention must be given to the release of certain radioactive gases. Among these are 12.26-yr tritium ($_1^3$H), which is a product of the occasional fission into three particles, 8.05-day iodine (I-131), and 10.76-yr krypton (Kr-85). The hazard associated with tritium is rather small because of the large MPC value, and the iodine concentration is greatly reduced by holding periods for fuel and process solutions. The long-lived krypton is a problem because it is a noble gas and thus cannot be reacted chemically for storage purposes. It may be disposed of in two ways—release to the atmosphere from tall stacks, with subsequent dilution by the atmosphere, or absorption of the chemically inert element in porous media such as charcoal operated at very low temperatures.

We can readily estimate the amount of krypton that must be handled. Since its half-life is long, the amount in $\frac{1}{3}$ of the fuel in a reactor operated for 3 yr is the same as that produced in the whole reactor in 1 yr. During the operation of a 3000 MWt reactor for 1 yr the number of fission events is

$$(3 \times 10^9 \, \text{W})\left(3.2 \times 10^{10} \frac{\text{fissions}}{\text{W-sec}}\right)\left(3.16 \times 10^7 \frac{\text{sec}}{\text{yr}}\right) = 3 \times 10^{27} \, \text{fissions/yr.}$$

Now the yield of Kr-85 in fission is around 0.3%, and thus the number of atoms produced is $N = 9 \times 10^{24}$. The decay constant is

$$\lambda = \frac{0.693}{t_H} = \frac{0.693}{(10.76)(3.16 \times 10^7)} = 2.0 \times 10^{-9} \, \text{sec}^{-1}.$$

Then the activity is

$$A = N\lambda = (9 \times 10^{24})(2 \times 10^{-9}) = 1.8 \times 10^{16} \, d/sec$$

or

$$A = \frac{1.8 \times 10^{16} \, d/sec}{3.7 \times 10^{10} \, d/sec\text{-}Ci} = 5 \times 10^{5} \, Ci.$$

It has been estimated that the total dosage to the world population resulting from complete release to the atmosphere of all Kr-85 from reactors would reach about 1 mrem/yr by the year 2000. Though this is a small figure, it is close enough to the 5 mrem/yr limit that steps will probably be taken to collect the gas and hold it for decay.

21.3 WASTE STORAGE AND DISPOSAL

After the uranium and plutonium are recovered, there remains a large volume of solution containing high-level fission products. Some of these such as strontium-90, cesium-137, and promethium-147 are valuable for industrial or medical applications, as will be described in Chapter 22. The total activity of such isotopes greatly exceeds the demand, however, and thus the bulk of the fission products have no presently known practical value. They must be stored or disposed of in such a way as to prevent exposure to the public.

Many ideas on what to do with these wastes have been advanced over the years since nuclear reactors started operating. For expediency, early byproducts of the plutonium production plants at Hanford, Washington were stored in solution in large underground tanks. Some consideration was given to the possibility of sealing concentrated wastes in concrete and dumping the containers in the ocean, but this concept was abandoned, since there was a chance of eventual erosion of the vessels, with the escape of long-lived radioactive materials.

A distinction must be made between *storage*, which implies that the material can be retrieved, and *disposal*, in which little or no future attention is required. A natural question arises—how long must wastes be retained? If absolute protection of the public from exposure is demanded, it is necessary to provide facilities that are perfect and truly permanent. The dominant isotope is Pu-239, which will be present in small amounts in wastes, even if the recovery is as high as 99.5%. The maximum permissible body burden for Pu-239 for occupational workers, as recommended by ICRP, is only 0.6 microgram, corresponding to an MPC in soluble form of $1 \times 10^{-4} \, \mu Ci/cm^3$ in water and $2 \times 10^{-12} \, \mu Ci/cm^3$ in air. It is fortunate that there are several obstacles to ingestion of plutonium by human beings, including low solubility, low uptake by plants, and small absorption by the body.

For storage that assures a high degree of safety, it is necessary to plan for very rigid control for many hundreds of years, during which the main fission products to consider are Sr-90 (27.7 yr) and Cs-137 (30 yr). In a period as long as 500 yr, which is 18 half-lives for strontium, the fraction that remains is still high, $2^{-18} = 6 \times 10^{-5}$.

Let us calculate the amount of Sr-90 produced per year from a 3000-MWt plant, yielding 3×10^{27} fissions. With a fission yield of 5.8%, and a decay constant $\lambda = 7.9 \times 10^{-10}$ sec^{-1}, the activity initially is found to be 3.7 megacuries. It is reasonable to assume that the concentrated liquid wastes from fuel reprocessing amount to 100 gal per 10,000 MWd. In one year, the reactor energy is 1.1×10^6 MWd, and thus the liquid volume is $(11,000 \text{ gal})(3785 \text{ cm}^3/\text{gal}) = 4.16 \times 10^7 \text{ cm}^3$. Thus the activity of Sr-90 is about 0.09 Ci/cm^3. This is to be compared with the MPC for Sr-90 for drinking water, which is $3 \times 10^{-7} \mu$Ci/cm^3. The ratio between these figures is 3×10^{11}, implying that the amount of leakage from a storage tank holding one year's production of Sr-90 to any source of public water must be restricted to essentially zero.

Studies have been made of the various means for achieving permanent disposal, such that little or no surveillance would be needed. One scheme advanced involved fusing the wastes with other solids, sealing the mixture in a container, and placing the capsules in underground salt beds. Some 10,000 mi^2 of such mines are available in the State of Kansas. The logic used was that the existence of the salt deposits implied freedom from access by water for millions of years. Thus one would expect future protection against attack of the containers by water and transfer of the contents by underground water flow. However, the discovery that there had been exploratory drilling for oil in the region being considered for disposal led to doubts as to the integrity of the salt bed method, and continued investigation has been undertaken.

Another suggested method involves storage very deep beneath the earth's surface, in cavities that are far below the level where water streams are found. It is possible that thermonuclear explosions could create the storage space required. The economic feasibility of sending concentrated wastes into space, preferably in a trajectory that would deposit the material in the sun, is being investigated, but the method will probably prove to be impractical.

For waste storage until an assured disposal system is developed, concrete and steel bunkers above ground will be employed, with continuous monitoring to discover leaks. It is fortunate that the very radiation that makes the nuclear wastes hazardous provides an extremely sensitive means for detecting escaping flow of radioactive fluids.

21.4 SUMMARY

Fission products constitute a waste from the operation of nuclear power plants. Although their volume is small, their radioactivity is extremely high, requiring that great care be exercised in the operations of fuel transfer, storage for decay, transportation, and reprocessing. Shipment is effected in casks that will withstand severe accidents. Chemical processing to remove the useful uranium and plutonium is followed by storage of fission products indefinitely, because of the very long half-lives of isotopes such as Pu-239, Sr-90, and Cs-137. Various storage and disposal methods have been considered, including above-ground in tanks and underground in salt deposits.

21.5 PROBLEMS

21.1. A batch of radioactive waste from a processing plant contains the following isotopes:

Isotope	Half-life	Fission yield, %
I-131	8 days	2.9
Ce-141	33 days	6
Ce-144	284 days	6.1
Cs-137	30 yr	5.9
I-129	1.7×10^7 yr	1

Letting the initial activity at $t = 0$ be proportional to λ and the fission yield, plot on semilog paper the activity of each for times ranging from 0 to 100 yr. Form the total and identify which isotope dominates at various times.

21.2. Traces of plutonium remain in certain waste solutions. If the initial concentration of Pu-239 in water were 100 parts per million ($\mu g/g$), find how much of the water would have to be evaporated to make the solution critical, neglecting neutron leakage as if the container were very large. Note: for H, $\sigma_a = 0.332$; for Pu, $\sigma_f = 742$, $\sigma_a = 1011$, $\nu = 2.87$.

21.3. If the maximum permissible concentration of Kr-85 in air is $1.5 \times 10^{-9} \mu Ci/cm^3$, and the yearly reactor production rate is 5×10^5 Ci, what is a safe diluent air volume flow rate (in cm^3/sec and ft^3/min) at the exit of the stack? Discuss the implications of these numbers in terms of protection of the public.

22

Beneficial Uses of Isotopes

Many important economic and social benefits are derived from the use of isotopes and radiation. The discoveries of modern nuclear physics have led to new ways to observe and measure physical, chemical, and biological processes, providing the strengthened understanding so necessary for man's survival and progress. The ability to isolate and identify isotopes gives additional versatility, supplementing techniques involving electrical, optical, and mechanical devices.

Special isotopes of an element are distinguishable and thus traceable by virtue of their unique weight or their radioactivity, while behaving chemically as the other isotopes of the element. Thus it is possible to measure amounts of the element or its compounds and trace movement and reactions.

When one considers the thousands of stable and radioactive isotopes available and the many fields of science and technology that require knowledge of process details, it is clear that a catalog of possible isotope uses would be voluminous. We shall be able here only to compare the merits of stable and radioactive species, to describe some of the special techniques, and to mention a few interesting or important applications of isotopes.

22.1 STABLE AND RADIOACTIVE ISOTOPES

Stable isotopes, as their name suggests, do not undergo radioactive decay. Most of the isotopes found in nature are in this category and appear in the element as a mixture. The principal methods of separation according to isotopic mass are electromagnetic, as in the large-scale mass spectrograph; and thermal-mechanical, as in the distillation or gaseous diffusion processes. Important examples are isotopes of elements involved in biological processes, e.g., deuterium and oxygen-18. The main advantages of stable isotopes are the absence of radiation effects in the specimens under study, the availability of an isotope of a chemical for which a radioactive species would not be suitable, and freedom from concern with speed in making measurements, since the isotope does not decay in time. Their disadvantage is the difficulty of detection.

Radioactive isotopes, or radioisotopes, are available with a great variety of half-lives, types of radiation, and energy. They come from three main sources—charged particle reactions in an accelerator, neutron bombardment in a reactor, and separated fission products. For example, the stable isotope of iodine is I-127; bombardment of tellurium-124 by deuterons yields a neutron and I-125 of half-life 60 days; absorption of neutrons in I-127 gives I-128, half-life 25 min; one of the iodine isotopes from fission is I-131, half-life 8 days. The main advantages of using radioisotopes are ease of detection of their presence through the emanations, and the uniqueness of the identifying half-lives and radiation properties. We shall now describe several special methods involving radioisotopes and illustrate their use.

22.2 TRACER TECHNIQUES

The tracer method consists of the introduction of a small amount of an isotope and the observation of its progress as time goes on. For instance, the best way to apply fertilizer containing phosphorus to a plant may be found by including minute amounts of the radioisotope phosphorus-32, half-life 14.2 days, emitting 1.7 MeV beta particles. Measurements of the radiation at various times and locations in the plant by a detector or photographic film provides accurate information on the rate of phosphorus intake and deposition. Similarly, circulation of blood in the human body can be traced by the injection of a harmless solution of radioactive sodium, Na-24, 14.9-hr half-life. For purposes of medical diagnosis, it is desirable to administer enough radioactive material to provide the needed data, but not so much that the patient is harmed.

The flow rate of many materials can be found by watching the passage of admixed radioisotopes. The concept is the same for flows as diverse as blood in the body, oil in a pipeline, or pollution discharged into a river. As sketched in Fig. 22.1, a small amount of radioactive material is injected at

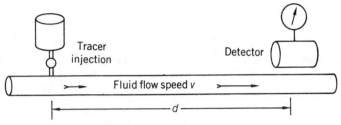

Fig. 22.1. Tracer measurement of flow rate.

a point, it is carried along by the stream, and its passage at a distance d away at time t is noted. In the simplest situation, the average fluid speed is d/t. It is clear that the half-life of the tracer must be long enough for detectable amounts to be present at the point of observation but not so long that the fluid remains contaminated by radioactive material.

In many tracer measurements for biological or engineering purposes, the effect of removal of the isotope by other means besides radioactive decay must be considered. Suppose, as in Fig. 22.2, that liquid flows in and out of a tank of volume V (cm^3) at a rate v (cm^3/sec). A tracer of initial amount N_0 atoms is injected and assumed to be uniformly mixed with the contents. Each second, the fraction of fluid (and isotope) removed from the tank is v/V, which serves as a flow decay constant λ_f for the isotope. If radioactive decay were small, the counting rate from a detector would decrease with time as $e^{-\lambda_f t}$. From this trend, one can deduce either the speed of flow or volume of fluid, if the other quantity is known. If both radioactive decay and flow decay occur, the exponential formula may also be used but with the effective decay constant $\lambda_e = \lambda + \lambda_f$. The composite effective half-life then can be found from the

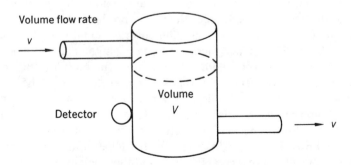

Fig. 22.2. Flow decay.

relation

$$\frac{1}{(t_H)_e} = \frac{1}{t_H} + \frac{1}{(t_H)_f}.$$

This logic applies equally well to the injection of a radioactive substance to an organism. The "biological half-life" takes the place of the flow half-life.

22.3 NEUTRON ACTIVATION ANALYSIS

This is an analytical method that will reveal the presence and amount of minute impurities. A sample of material that may contain traces of a certain element is irradiated with neutrons, as in a reactor. The gamma rays emitted by the product radioisotope have unique energies and relative intensities, in analogy to spectral lines from a luminous gas. Measurements and interpretation of the gamma ray spectra, using data from standard samples for comparison, provide information on the amount of the original impurity.

Let us consider a practical example. Reactor design engineers may be concerned with the possibility that some stainless steel to be used in moving parts in a reactor contains traces of cobalt, which would yield undesirable long-lived activity if exposed to neutrons. To check on this possibility, a small sample of the stainless steel is irradiated in a test reactor to produce Co-60, and gamma radiation from the Co-60 is compared with that of a piece known to contain the radioactive isotope. The "unknown" is placed in a well-shielded container with a scintillation detector of the type described in Chapter 11. Gamma rays from the decay of the 5.26-yr Co-60 give rise to electrons by photoelectric absorption, Compton scattering, and pair production. The electrons then produce light pulses and thus electrical signals in approximate proportion to the energy of the gamma rays. If all of the pulses produced by gamma rays of a single energy were equal in height, the observed counting rate would be as shown in Fig. 22.3a, since there are two cobalt gammas of energy 1.17 and 1.33 MeV. However, there are variations in both the number of photons resulting from the collisions in the detector and in the electron production in the photomultiplier. Also, because of photon scattering the energy of some gamma rays is reduced, and the interaction with the crystal gives pulses of lower energy. Thus actual measurements show a spread in energy as shown in Fig. 22.3b. Even so, the location of the photopeaks clearly indicates the presence of Co-60, and their height provides a basis for estimating the amount present. Modern electronic circuits can process

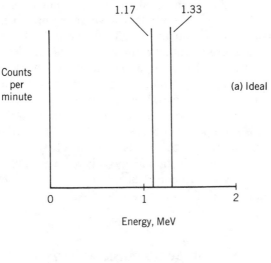

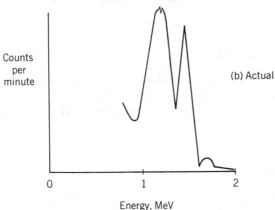

Fig. 22.3.   Analysis of gamma rays from Co-60.

a large amount of data at one time. The multichannel analyzer accepts counts due to photons of all energy and displays the whole spectrum graphically.

When neutron activation analysis is applied to a mixture of materials, it is necessary after irradiation to allow time to elapse for the decay of certain isotopes whose radiation would "compete" with that of the isotope of interest. In some cases, prior chemical separation is required to eliminate undesirable radiation.

The activation analysis method is of particular value for the identification of chemical elements that have an isotope of high neutron absorption cross section, and for which the products yield a suitable radiation type and energy. Not all elements meet these specifications, of course, which means that activation analysis supplements other techniques. For example, neutron absorption in the naturally occurring isotopes of carbon, hydrogen, oxygen, and nitrogen produces stable isotopes. This is fortunate, however, in that organic materials including biological tissue are composed of those very elements, and the absence of competing radiation makes the measurement of trace contaminants easier. The sensitivity of activation analysis is remarkably high for many elements, it being possible to detect quantities as low as a millionth of a gram in 76 elements, a billionth of a gram in 53, or even as low as a trillionth in 11.

22.4 APPLICATIONS OF ACTIVATION ANALYSIS

A few of the many practical uses of the method are cited.

(a) Textile Manufacturing

In the production of synthetic fibers, certain chemicals such as fluorine are applied to improve textile characteristics, such as the ability to repel water or stains. Activation analysis is used to check on inferior imitations, by comparison of the content of fluorine or other deliberately added trace elements.

(b) Petroleum Processing

The "cracking" process for refining oil involves an expensive catalyst that is easily poisoned by small amounts of vanadium, which is a natural constituent of crude oil. Activation analysis provides a means for verifying the effectiveness of the initial distillation of the oil.

(c) Crime Investigation

The process of connecting a suspect with a crime involves physical evidence that often can be accurately obtained by activation analysis. Examples are: the comparison of paint flakes found at the scene of an automobile accident with paint from a hit-and-run driver's car; the determination of the geographical source of drugs such as opium by comparison of trace element content with that of soils in which the poppy plant can be grown; the verification of the source of copper telephone wire believed to be stolen, by taking account of variations in the chemical content of wire from different manufacturers; measurement of the amounts of barium or antimony on the hands of suspects or victims in gunshot cases where it is not clear whether murder or suicide is involved; the identification of the origin of bullets by their unique antimony content; and tests for poison in a victim's body. The classic example of the latter is the verification of the hypothesis that Napoleon was poisoned, by use of activation analysis measurements of the content of arsenic in samples of his hair.

(d) Authentication of Paintings

The probable age of a work of art can be found by testing a small speck of paint. Over the centuries, the proportions of elements such as chromium and zinc used in paint have changed, and forgeries of the work of old masters can thus be detected.

(e) Diagnosis of Disease

Promising medical applications still under investigation include the measurement of sodium content in children's fingernails in the diagnosis of cystic fibrosis, and the accurate measurement of normal and abnormal content of some 50 trace elements in the blood, as indicators of specific diseases.

(f) Pesticide Investigation

The amounts of residues of pesticides such as DDT or methyl bromide in crops, foods, and animals are found by analysis of the bromine and chlorine content.

(g) Mercury in the Environment

The heavy element mercury is a serious poison for animals and human beings even at low concentrations. It appears in rivers as a result of discharge of certain manufacturing process wastes. By the use of activation analysis, which can measure down to around $\frac{1}{20}$ of a part per million of mercury, the amount of mercury contamination in water or tissues of fish or land animals can be determined, helping establish the chain by which it is transferred to biological species.

22.5 GAGING

Some physical properties of materials are difficult to ascertain by ordinary methods, but can be measured readily by observing how radiation interacts with the substance. For example, the thickness of a thin layer of plastic or paper can be found by measuring the number of beta particles from a radioactive source that are transmitted. The separated fission product isotopes Sr-90 (27.7 yr, 0.54 MeV β particle) and Cs-137 (26.6 yr, 0.52 MeV β particle) are widely used for such gaging.

The density of a liquid flowing in a pipe can be measured externally by detection of the gamma rays that pass through the substance. The liquid in the pipe acts as a "shield" for the radiation, with attenuation dependent on macroscopic cross section and thus particle number density, as discussed in Chapter 19.

The level of liquid in an opaque container can be measured readily without the need for sight glasses or electrical contacts. A detector outside the vessel measures the radiation from a radioactive source mounted on a float in the liquid.

The moisture content of soils can be estimated by study of the neutrons slowed by hydrogen. In the neutron moisture gage, a source consisting of a mixture of an alpha particle emitter, e.g., Pu-239, and beryllium Be-9 provides fast neutrons by the (α, n) reaction. The flux of thermal neutrons measured by a BF_3 counter provides data on the water content.

Several nuclear techniques are employed in the petroleum industry. In the drilling of wells, the "logging" process involves the study of geological features. One method consists of measurement of natural gamma radiation. When the detector is moved from a region of natural radioactive rock to one containing oil or other liquid, the signal is reduced. The neutron moisture gage is also adapted to determine the presence of oil, which contains hydrogen. Neutron activation analysis of chemical composition is performed by lowering a neutron source and a gamma ray detector into the well.

22.6 DATING

There would appear to be no relation between nuclear energy and the humanities such as history, archeology, and anthropology. There are, however, several interesting examples in which nuclear methods establish dates of events. The carbon dating technique is being used regularly to determine the age of ancient artifacts. The technique is based on the fact that carbon-14 is and has been produced by cosmic rays in the atmosphere (a neutron reaction with nitrogen). Plants take up CO_2 and deposit C-14, while animals eat the plants. At the death of either, the supply of radiocarbon obviously stops and that present decays, with half-life 5730 yr. By measurement of the radioactivity, the age within about 50 yr can be found. This method was used to determine the age of the Dead Sea Scrolls, as about 2000 yr, making measurements on the linen made from flax; to date the documents at Stonehenge in England, using pieces of charcoal; and to verify that prehistoric peoples lived in the United States, as long ago as 9000 yr, from the C-14 content of rope sandals discovered in an Oregon cave.

The age of minerals in the earth, in meteorites, or on the moon can be obtained by a comparison of their uranium and lead contents. The method is based on the fact that Pb-206 is the final product of the decay chain starting with U-238, half-life 4.51×10^9 yr. Thus the number of lead atoms now present is equal to the loss in uranium atoms, i.e.,

$$N_{Pb} = (N_U)_0 - N_U,$$

where

$$N_U = (N_U)_0 e^{-\lambda t}.$$

Elimination of the original number of uranium atoms $(N_U)_0$ from these two formulas gives a relation between time and the ratio N_{Pb}/N_U. The latest value of the age of the earth obtained by this method is 4.55 billion years.

For the determination of ages ranging from 50,000 to a few million years, an argon method can be employed. It is based on the fact that the potassium isotope K-40 (half-life 1.62×10^9 yr) crystallizes in materials of volcanic origin and decays into the stable argon isotope Ar-40. The technique is of particular interest in establishing the date of the first appearance of man.

22.7 SUMMARY

Stable and radioactive isotopes have great utility for measurements of properties and processes in many fields. The tracer technique provides information on flow of fluids, neutron activation analysis tells the amounts of minute impurities in a great variety of applications, and isotope gages measure thickness, density, and moisture content. Accurate dating of ancient specimens is made possible by measurements of C-14, of argon from decay of K-40, and of the ratio of lead to uranium.

22.8 PROBLEMS

22.1. A radioisotope is to be selected to provide the signal for arrival of a new grade of oil in a 500-mile-long pipe line, in which the fluid speed is 3 ft/sec. Some of the candidates are:

Isotope	Half-life	Particle, energy (MeV)
Na-24	14.96 h	β 1.389; γ 1.369, 2.754
S-35	87.9 d	β 0.167
Co-60	5.263 y	β 0.314; γ 1.173, 1.332
Fe-59	45 d	β 1.57; γ 1.095, 1.292

Which would you pick? On what basis did you eliminate the others?

22.2. The radioisotope F-18, half-life 110 min, is used for tumor diagnosis. It is produced by bombarding lithium carbonate (Li_2CO_3) with neutrons, using tritium as an intermediate particle. Deduce the two nuclear reactions.

22.3. The range of beta particles of energy 0.53 MeV in metals is 170 mg/cm^2. What is the maximum thickness of aluminum sheet, density 2.7 g/cm^3, that would be practical to measure with a Sr-90 or Cs-137 gage?

22.4. The amount of environmental pollution by mercury is to be measured using neutron activation analysis. Neutron absorption in the mercury isotope Hg-196, present with 0.146% abundance, activation cross section 3080 barns, produces the radioactive species Hg-197, half-life 65 hr. The smallest activity for which the resulting photons can be accurately analyzed in a river water sample is 10 d/sec. If a reactor neutron flux of 10^{12} cm^{-2}-sec is available, how long an irradiation is required to be able to measure mercury contamination of 20 ppm (μg/g) in a 4 milliliter water test sample?

22.5. The ratio of numbers of atoms of lead and natural uranium in a certain moon rock is found to be 0.05. What is the age of the sample?

22.6. The activity of C-14 in a wooden figure found in a cave is $\frac{3}{4}$ of its modern value. Estimate the date the figure was carved.

22.7. Examine the possibility of adapting the uranium–lead dating analysis to the potassium–argon method. What would be the ratio of Ar-40 to K-40 if a deposit were 1 million years old?

22.8. The age of minerals containing rubidium can be found from the ratio of radioactive Rb-87 to its daughter Sr-87. Develop a formula relating this ratio to time.

23

Applications of Radiation

The nuclear methods just described emphasized the detection and measurement of radiation as a means to identify the emitters. We now turn to the beneficial effects of radiation from various sources—X-ray machines, charged particle accelerators, nuclear reactors, and radioisotope sources. The penetrating radiations are electromagnetic, electrons or other charged particles, and neutrons. Applications are cited in industry, medicine, agriculture, and space exploration, with the examples selected for importance and interest.

23.1 RADIOGRAPHY

The oldest and most familiar beneficial use of radiation is for medical diagnosis by X-rays. These consist of high-frequency electromagnetic radiation produced by electron bombardment of a heavy-metal target. As is well known, X-rays penetrate body tissue to different degrees dependent on material density, and shadows of bones and other dense material appear on the photographic film. The term radiography includes the investigation of internal composition of living organisms or objects in general, using X-rays, gamma rays, or neutrons.

For both medical and industrial use, the isotope Co-60, produced from stable Co-59 by neutron absorption, is an important alternate to the X-ray tube. Co-60 provides gamma radiation of energies 1.17 and 1.33 MeV, which are especially useful for examination of flaws in metals. Internal cracks, defects in welds, and nonmetallic inclusions are revealed by scanning with a cobalt radiographic unit. Advantages include small size and portability, and freedom from the requirement of an electrical power supply. The half-life of 5.26 yr permits use of the cobalt source for a long time without refueling. On the other hand, the energy of rays is fixed and the intensity cannot be varied, as is possible with the X-ray machine. For radiography of thin specimens, the isotope iridium-192 is convenient. Its half-life is 74.2 days and the photon energies around 0.4 MeV.

Neutron radiography serves as a complement to gamma ray radiography where the materials are insensitive to photons but are rich in hydrogen, e.g., plastics and rubber. The neutron sources are typically antimony–beryllium, in which the gamma rays from Sb-124, half-life 60.4 days, cause a (γ, n) reaction in Be-9. A promising new source is californium-252, an artificial isotope (element 98) produced by successive neutron bombardment of plutonium in a reactor. Although Cf-252 usually decays by alpha particle emission, about 3% of the time it undergoes spontaneous fission, emitting around 3.5 neutrons. An extremely small amount of the isotope thus serves as a copious source of neutrons.

23.2 MEDICAL APPLICATIONS

A rapid growth in the use of radiation for medical therapy has taken place in recent years, with millions of treatments of patients administered annually. The radiation can come from teletherapy units, in which the source is at some distance from the patients, or from isotopes in sealed containers implanted in the body, or from injected or ingested solutions containing isotopes.

Doses of radiation are often found to be effective in the treatment of certain diseases such as cancer. Over the years, X-rays have traditionally been used, but it has been found that the penetrating Co-60 gamma rays permit higher doses to tissue deep in the body, with a minimum of skin reaction. The cobalt equipment requires no expensive electrical maintenance.

Considerable success in treatment of abnormal pituitary glands has been obtained by irradiations with charged particles from an accelerator, and slow neutron irradiation of tumors injected with boron solution has been found beneficial in some cases. Diseases such as leukemia and hyperthyroidism respond to treatment by radiation from radioisotopes of phosphorus and iodine, respectively.

The prevention of infection in surgery requires use of sutures that have been sterilized. Conventional inefficient batch processing with chemicals and heat has been largely replaced by mass-production methods involving bombardment of packaged sutures by radiation. Electron beams from a Van de Graaff accelerator or gamma rays from a Co-60 source can be used for this purpose.

23.3 FIBER IMPROVEMENT

Various properties of polymers such as polyethylene can be changed by electron or gamma ray irradiation. The original material consists of very long parallel chains of molecules, and the radiation causes the chains to be connected, a process called cross-linking. Irradiated polyethylene has better resistance to heat and serves as excellent electrical insulating coating for wires. Fabrics can be made soil-resistant by a process of radiation bonding of a suitable polymer to a fiber base.

23.4 SYNTHESIS OF CHEMICALS

Certain chemical reactions can be initiated using high-energy gamma radiation. Many of these reactions are feasible in a laboratory, but relatively few processes can be made economical. An exception is the production of ethyl bromide (CH_3CH_2Br), a volatile organic liquid used as an intermediate compound in the synthesis of organic materials. Gamma radiation from a Co-60 source has the effect of a catalyst in the combination of hydrogen bromide (HBr) and ethylene (CH_2CH_2). Gammas are found to be superior as catalysts to chemicals, or application of ultraviolet light, or electron bombardment. Millions of pounds of ethyl bromide are produced annually by this unique process.

The commercially important chemical polyethylene is also produced by cobalt gamma ray bombardment of ethylene.

23.5 WOOD PLASTIC PROCESSING

There is a large commercial demand for a new type of wood flooring produced by gamma irradiation. Wood is soaked with a plastic and passed through a beam of gamma rays, which changes the molecular structure of the plastic and leaves a surface that cannot be scratched or burned. The appearance of the product is unchanged, but the material is extremely durable, making the wood especially useful for public areas such as lobbies of airport terminals. The extra cost of processing is justified by the long useful life of the material, and millions of square feet are manufactured each year. Similar techniques of irradiation are used in the preparation of architectural tiles with much improved wear and strength characteristics.

23.6 RADIATION PRESERVATION OF FOOD

As early as 1953, studies were under way on the feasibility of large-scale processing of food by irradiation with gamma rays. It has been shown repeatedly that dramatic improvements in the shelf life of foods are effected, of the order of month, and there is no evidence that radioactivity is induced by gamma ray bombardment. There are two unresolved problems, however—in some foods, subtle changes in taste are noted, some of which are disagreeable, and there is concern that the radiation may induce chemical changes that render the food unsafe.

Among examples of produce that have been tested are potatoes (to inhibit sprouting), bacon, wheat (for disinfestation), strawberries (to prevent decay and rot), and fish. Research is under way on the organism that produces botulism in fish products. A jurisdictional problem involving federal agencies—the Food and Drug Administration, the Atomic Energy Commission, and the Army—has unfortunately slowed progress in this important area.

23.7 CROP MUTATIONS

The science of crop breeding involves the selection of unusual plants and crossing them to obtain permanent and reproducible hybrids with desired properties. Crops with high yield, resistance to disease and adaptability to new environments have been obtained by such genetic studies. However, the process can be accelerated by the application of radiation, such as charged particles, X-rays, gamma rays, and neutrons. Desirable mutations are induced by irradiation of seeds or by the removal of cuttings from irradiated trees. Success in obtaining new strains has been achieved in beans, oats, barley, peanuts, and many types of ornamental flowers and plants.

The improvement of food production by crop mutations is of especial importance in the problem of an expanding world population, in terms of both a higher yield of crops and a higher nutritional value. For example, a new strain of mutant rice has been developed that contains twice as much protein as conventional varieties.

23.8 INSECT CONTROL

To eradicate certain insect pests, the so-called sterile male technique has been successfully applied. The procedure for eradication consisted of laboratory breeding of large numbers of male insects, sterilizing them with gamma radiation, and releasing them for mating in the infested area. Competition of the sterile males with normal males causes a rapid reduction in population over a two-year period. The most dramatic example is the elimination in the South and Southwest of the screw worm fly, a pest that had caused millions of dollars of damage to livestock. The fly lays eggs in wounds and the larva kills the animal.

The method also has promise for eradicating the tsetse fly, the carrier of the disease sleeping sickness, which is very prevalent in Africa. Many millions of acres of land are now uninhabitable because of the presence of the insect. The main problem to be solved is the technique for rearing and sterilizing the flies in adequate numbers.

23.9 ISOTOPIC POWER GENERATORS

The long half-life of certain radioisotopes make them very suitable in the construction of light, compact, and reliable power sources, especially for remote locations. One of the most important of such isotopes is Pu-238, half-life 86.4 yr, which emits alpha particles of 5.5 MeV energy. The isotope $^{238}_{94}Pu$ is produced by reactor neutron irradiation of the almost-stable isotope $^{237}_{93}Np$ (2.14×10^{6}-yr half-life). The latter is a decay product of $^{237}_{92}U$, a 6.75-day beta emitter that arises from neutron capture in $^{236}_{92}U$ or by (n, 2n) and (γ, n) reactions with $^{238}_{92}U$. The high-energy alpha particles and the relatively short half-life of Pu-238 give the isotope the favorable power to weight ratio 0.57 W/g and a high specific activity of 17 Ci/g.

In the Apollo-12 mission to the moon in November 1969, a group of scientific instruments called ALSEP (Apollo Lunar Surface Experimental Package) was set up to measure magnetic fields, dust, the solar wind, ions, and earthquake activity. It was powered by a 74-watt source using Pu-238. The heat source of 1480 W required 44,500 Ci corresponding to about 2.6 kg of Pu-238.

The isotopic electrical power generator for Apollo-12 is shown schematically in Fig. 23.1. Between the rod of plutonium dioxide and the beryllium case are located 442 thermoelectric couples composed of

Radioisotope fuel capsule

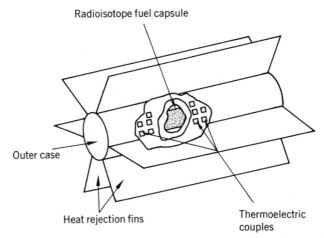

Outer case

Heat rejection fins

Thermoelectric couples

Fig. 23.1. Isotopic electrical power generator. (SNAP-27 used in Apollo-12 mission.)

lead-telluride, connected to yield 74 W at 16 V. The efficiency is thus around 5%. The isotopic capsule operates at 1350°F; hot and cold junctions are at 1100°F and 525°F, respectively. The generator weighs 45 lb, whereas batteries to do the same job would weigh 4000 lb. The power supply, like the popular watch, "keeps on ticking," even though the temperature swing on the moon is from −280°F to 250°F. The fuel was transported on the Apollo-12 flight in a container designed for "intact impact" should the mission be aborted, and for resistance to the intense heat experienced on reentry into the atmosphere. The generator was then assembled on the lunar surface by the astronauts, connected to the ALSEP, and deployed for measurements. An isotopic power generator will also be used as a power source with two-year life expectancy in the long-range unmanned trips to Jupiter and Mars during the 1970s.

A very promising medical spinoff of the development of the isotopic generator developed for the space program is the heart pacemaker, which provides small electrical impulses to regulate heartbeat. Pacemakers of a few hundred microwatts, powered by small quantities of Pu-238, will last for many years and are preferable to those powered by batteries, requiring frequent replacement by surgical operation. Such long life makes the isotopic source attractive for brain pacemakers, which stop epileptic seizures.

Another future use is suggested by the news of a successful implantation of an auxiliary heart pump that consists of a mechanically operated device to parallel the patient's own heart. The technique eliminates the need for a heart transplant, which has shown to be of limited success. The next step would be to provide the power for the new device by isotopic means, and one can anticipate artificial organs employing compact and reliable isotopic generators.

23.10 SUMMARY

Many applications of penetrating radiations are found in industry, medicine, agriculture, and space exploration. Radiography uses X-rays, Co-60 gamma rays, and neutrons to inspect the body or bulk materials. Various radiations are applied for therapeutic benefit, improvement of fibers, synthesis of chemicals, the production of wood plastic, food preservation, beneficial crop mutations, and elimination of certain insect pests. Isotopic generators provide heat and electrical energy for space experiments and are available for heart pacemakers.

23.11 PROBLEMS

23.1. The half-life of Cf-252 for spontaneous fission is 85 yr. Assuming that it releases three neutrons per fission, how much of the isotope in micrograms is needed to provide a source of strength of 10^7 neutrons/sec? What would be the diameter of the source in the form of a sphere if the Cf-252 had a density as pure metal of 20 g/cm^3?

23.2. Three different isotopic sources are to be used in radiography of steel in ships as follows:

Isotope	Half-life	Gamma energy (MeV)
Co-60	5.26 yr	1.25 (ave.)
Ir-192	74.2 days	0.4 (ave.)
Cs-137	30 yr	0.66

Which isotope would be best for insertion in pipes of small diameter and wall thickness? For finding flaws in large castings? For more permanent installations? Explain.

23.3. A cobalt source is to be selected for irradiation of potatoes to inhibit sprouting. What strength in curies is needed to process 500,000 lb of potatoes per day, providing a dose of 10,000 rad? Note that two gammas totaling around 2.5 MeV energy are emitted by Co-60. What amount of isotopic power is involved? Discuss the practicality of absorbing all of the gamma energy in the potatoes.

23.4. (a) Verify that Pu-238, half-life 86.4 yr, alpha energy 5.5 MeV, yields an activity of 17 Ci/g and a power of 0.57 W/g.

(b) How much plutonium would be needed for a 200 μW heart pacemaker?

24

Nuclear Explosives

The intent of the present book is to emphasize the beneficial applications of nuclear energy, recognizing of course that the threat of warfare with nuclear weapons exists. There is one point of connection, however—the possibility of using the tremendous energy release of a nuclear explosive for peaceful purposes, being investigated in a program called "Plowshare."†

In this chapter we shall briefly describe the action of a nuclear device and its effects, and discuss applications of explosives that have been studied or tested, including large-scale excavations and the stimulation of release from the earth of oil, natural gas, and heat energy.

24.1 THE NUCLEAR EXPLOSION

Security of information on the detailed construction of nuclear weapons has been maintained, and only a qualitative description is available to the public. Figure 24.1 shows sketches from an article by Herbert F. York,‡ and his caption for the figure is transcribed verbatim:

†*Holy Bible*, Isaiah 2-4 and Micah 4-3, "And he shall judge among the nations, and shall rebuke many people: and they shall beat their swords into plowshares, and their spears into pruning hooks: nation shall not lift up sword against nation, neither shall they learn war any more."

‡Herbert F. York, "The Great Test-Ban Debate," *Scientific American*, November 1972, pages 15–23.

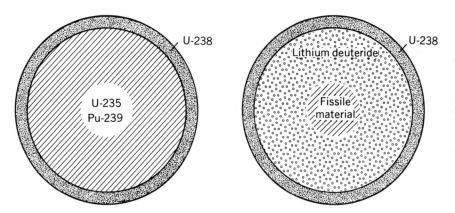

Fig. 24.1. Nuclear weapons. (From *The Great Test-Ban Debate* by Herbert F. York. Copyright © November 1972 by Scientific American, Inc. All rights reserved.)

Nuclear weapons are depicted schematically. At left a fission bomb is shown just as the nuclear explosion starts. A supercritical mass of fissile material has been rapidly assembled by means of high explosive, which is not shown, and a chain reaction is initiated by a neutron from a source near the mass. The uranium 238 reflects some neutrons and provides some additional inertia but does not otherwise participate importantly in the reaction. The drawing of a thermonuclear bomb (right) explains its known properties, although no detailed design has ever been made public. Fissile material in the central core explodes, producing much heat and many neutrons. The neutrons convert some of the lithium 6 surrounding the core to tritium, and the heat causes thermonuclear deuterium–deuterium and deuterium–tritium reactions to take place in the lithium deuteride. These reactions in turn produce neutrons having enough energy to cause the surrounding uranium 238 to fission, thus releasing still more energy and additional quantities of neutrons.

Thus we see that thermonuclear explosions are produced by combining the fission and fusion reactions. The fission of enriched uranium or plutonium provides the necessary energy to heat up a mixture of deuterium and tritium to the temperature of millions of degrees required for the fusion reaction to start, as also discussed in Chapter 8. The event occurs almost instantaneously (within a microsecond), with extremely high temperatures developed that vaporize and ionize all material in the immediate vicinity. High accompanying pressures (millions of atmospheres) can displace large amounts of material. The compactness and energy yield of a nuclear charge in comparison with those for a chemical charge such as dynamite or TNT (trinitrotoluene) is of great advantage.

24.2 UNDERGROUND ENGINEERING

The Plowshare program has yielded a great deal of information on the technology of underground explosions. In a typical test, a hole is drilled several thousand feet deep, the thermonuclear device is lowered to the bottom of the shaft, and the fission–fusion reaction is set off. The amount of energy release can be predetermined by the construction of the device. Suppose, for example, the total energy release is equivalent to that from 100 kilotons† of TNT. Of this, 1% might be fission energy, 99% fusion

Fig. 24.2. Underground cavity caused by fusion explosion. (Courtesy of Lawrence Research Laboratory, Livermore, and the United States Atomic Energy Commission.)

†The amount of energy from explosion of TNT varies somewhat, but by definition 1 ton of TNT corresponds to 10^9 cal of energy.

energy. Detonation produces a shock wave, consisting of material moving outward at uniform speed into the surroundings. Since the shock is composed of an ionic plasma at extremely high temperature, it vaporizes, melts, crushes, displaces, or cracks the rock as the energy is dissipated. A large cavity in the previously solid rock is produced. For example, a nuclear explosive equivalent to 300 kilotons of TNT buried to a depth of 1200 ft creates a spherical cavity of about 170 ft in diameter, largely filled with broken rock and gases at temperatures of several thousand degrees centigrade. Figure 24.2 shows the result of such an explosion.

The character of the cavity depends on the placement of the charge. It has been found that the volume of cavity produced is directly proportional to the energy release and varies inversely (approximately) with the weight of the column of rock above the point of detonation, i.e., the deeper the shot, the smaller the cavity. Figure 24.3 shows schematically the effect of a deep underground explosion. Extending vertically upward from the shot point is a column of cracked or broken rock. This space, called a "chimney," is often several times the diameter of the cavity.

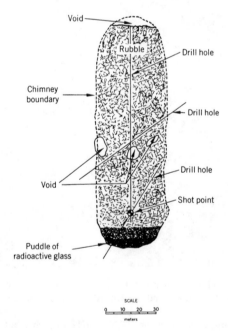

Fig. 24.3. Effect of underground nuclear explosion.

24.3 NATURAL GAS STIMULATION

One proposed use of underground nuclear explosives is for the stimulation of natural gas production. The mechanism for improvement is readily understood. In a conventional well, the gas flows from the slightly permeable rock of an underground reservoir to the well bore, which is about 6 in. in diameter. When a nuclear explosion takes place deep in the ground, the region of broken rock, many hundreds of feet in diameter, now becomes the new effective well bore. The collection area is thus multiplied by a large factor.

The two initial experiments aimed at stimulation of gas production were Project Gasbuggy, 29 kilotons at 4240 ft in New Mexico, 1967, and Project Rulison 42 kilotons at 8426 ft in Colorado, 1969. Continued measurements of gas pressure and flow showed an improvement in production.

Accompanying the explosions were radioactive products and the long term usefulness of the process depends on control of the amount of activity that gets into the gas. It appears that fission products such as iodine are trapped on rock surfaces, but that there are significant amounts of Kr-85 and tritium. The tritium is the isotope of greatest concern. However, it will appear only in the initial gas, which can be dispersed with dilution, or which can be burned in a controlled way to collect the tritium. Alternatively, an explosive based on fusion only could be used.

As the reserves of natural gas decline and the price increases, nuclear stimulation methods may become increasingly attractive. The gas could be pumped out or stored in the cavity for future use as needed. It has been estimated that a stimulated well would produce about 15 times that of a conventional well. Under these conditions, the process might be economically feasible, even though each 100-kiloton device costs around half a million dollars. It has been predicted that the United States reserves of around 2×10^{11} ft^3 could be doubled by successful application of the nuclear method to presently nonproductive gas basins.

24.4 NUCLEAR EXCAVATION

There are many potential uses for nuclear explosives in civil engineering works. Over the years since 1957 when the Plowshare program began, engineering studies have been made on the feasibility and economic benefit of projects such as a new canal between the Atlantic and Pacific oceans in Panama, Colombia, or Nicaragua, the excavation of harbors in

Alaska and Australia, cuts in California mountains to accommodate railroad and highway traffic, a canal to connect two rivers in the South, and the production of large quantities of rock for a dam in Idaho. Several of these were found to be technically possible, but for various reasons have not been carried out.

For nuclear excavation projects, charges are buried near the surface of the earth, and the explosions produce craters whose size and shape depend on the energy and placement of the explosive and on the geological formation. A great deal of knowledge is available on these effects. More information is needed however on the possibility and extent of damage by earthquakes induced by a series of nuclear explosions.

Although there are good indications that large-scale projects such as an interoceanic canal could be carried out without excessive release of radioactivity, there are some political concerns related to the Limited Test Ban Treaty of 1963. Among the provisions of this international agreement is the prohibition of transfer of radioactive debris across a nation's borders. There is much ambiguity as to whether this means hazardous amounts or merely detectable amounts.

24.5 EXTRACTION OF OIL, MINERALS, AND HEAT

Consideration has been given to the possibility of using nuclear explosives for a variety of other purposes.

(a) Large amounts of oil are present in oil shale, which contains solid hydrocarbons. It has been proposed to break up the shale rock by a nuclear explosion. Subsequently, air would be supplied to the cavity to maintain a fire that heats and decomposes the hydrocarbons to release liquid oil.

(b) The mining of ores such as copper is not economical by conventional means when the deposits are too deep. An explosion would create a cavity filled with ore-bearing rock, and acid would be pumped in to extract the mineral.

(c) The existence in the earth of large amounts of stored heat energy has long been recognized. In some locations, steam is readily available for generation of electricity. Nuclear explosives might be used to make such geothermal power available in many places. Where there is a high enough temperature of the rock formation, an explosion would break up the rock, and water would be piped in to be converted into steam, as sketched in Fig. 24.4.

In all such potential applications, the effect of radioactivity must be carefully considered. The slow decay of composite fission products has

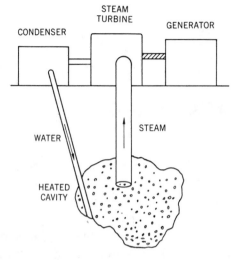

Fig. 24.4. Explosion-stimulated geothermal power.

been described earlier (Chapter 20). Among the fission products of primary concern are those of intermediate half-life such as Sr-90, Cs-137, and I-131, while the fusion reaction yields tritium and C-14.

24.6 SUMMARY

The tremendous energy release of thermonuclear explosions suggests many potential applications. An underground detonation produces a very large cavity filled with broken rock. Natural gas production has been shown to be stimulated by nuclear explosives, and part of our long-range energy needs may be met by their use. Many types of excavation are feasible—canals, cuts, tunnels, dams, harbors, and water reservoirs, but as yet no major projects have been accomplished. Other resources of the earth that might be tapped are oil, minerals such as copper, and heat energy. In all nuclear explosive applications, attention must be paid to the release of radioactivity and to seismic effects.

24.7 PROBLEM

24.1. A proposal is advanced to pipe the heat of fusion explosions from a cavity deep in the earth up to the surface. If no energy were lost, how often would a 100-kiloton device have to be set off in order to obtain 3000 MW of thermal power? What would be the cost of the explosives per year? How does this "fuel" cost compare with that from fission (see Chapter 13)?

25

Nuclear Propulsion

The idea of using nuclear power systems for transportation is attractive because of the very large energy released per pound of fuel. When trips of long duration or distance are required, propulsion by nuclear means is distinctly advantageous. Over the period since 1955, however, the only large-scale continued use of nuclear methods for propulsion has been for military naval vessels. Research and development projects toward nuclear-powered aircraft and space vehicles have been carried out, but were abandoned or suspended for various reasons. In this chapter we shall discuss the virtues of nuclear energy for propulsion, its limitations, its accomplishments, and prospects for the future.

25.1 ADVANTAGES AND LIMITATIONS OF NUCLEAR PROPULSION

To put the application of nuclear power for propulsion into perspective, let us review some of the existing types of transportation. For land, we have motorcycles, snowmobiles, automobiles, buses, trucks, and railroad trains; for water, we have speedboats, hovercraft, passenger liners, merchant ships, aircraft carriers, and submarines; for air, we have helicopters, airplanes, and for certain purposes balloons and dirigibles; for space, we have rockets.

For the smaller of such vehicles, nuclear power is unsuitable for three main reasons. First is the cost of the initial fuel to achieve criticality. The smallest amount of U-235 that can be regarded as feasible is 1 kg, which would require an expense of about ten thousand dollars for fuel alone. To this would be added the very high cost of the accompanying equipment. Second is the weight of shielding needed to protect the driver, occupants, and passersby. A 100-horsepower engine, corresponding to about 75 kW, would yield 2.5×10^{15} neutrons/sec. Using simple methods as described in Chapter 19, we find that a spherical water shield that would reduce the radiation to safe levels would have a radius about 8 ft, and a weight of some 70 tons. Third is the potential for release of radioactive fission products in the event of a collision that breaks open the reactor vessel. We note that the probability of accident is greater the smaller and more mobile the conveyance. Nuclear propulsion for any mode of transportation is feasible only if the reactor, its shield, energy conversion equipment, and protective barriers constitute a small fraction of the total weight of the vehicle. Studies have been made of reactor powered railroad engines, snow-trains, and hovercraft, but these applications are generally regarded as marginal at best. For any smaller vehicles that carry passengers or must operate in congested areas, nuclear power is not feasible.

An all-electric ground transportation system based on nuclear power is a very attractive concept. Automobiles and trucks would be powered by quickly replaceable electric batteries that would be charged at existing (but adapted) service stations, which receive their energy supply from a network of suitably located nuclear power plants. Passenger and freight trains and inter-urban mass-transit systems would be electrically powered from lines along the route. The advantages of such a system are obvious: the tremendous pollution of the environment by chemical fumes from the internal combustion engine would be eliminated, along with smoke and gas from fossil-fueled electrical power plants; our resources of gas, oil, and coal would be conserved for alternative beneficial uses; traffic safety would be enhanced by the probable lower speed attainable by the vehicles and by the elimination of the possibility of fire accompanying an accident. The approach is particularly attractive for urban areas where high speed is not needed and where pollution is greatest. For long trips, a combination of high-speed rail transport and inexpensive rental cars would provide optimum convenience.

An alternative is to use electricity to dissociate water and to store, transport, and burn the hydrogen gas just as is done for natural gas. When hydrogen gas is used in an internal combustion engine, pollution of the atmosphere is a minimum.

The airplane is the first in the scale of increasing vehicle size that would appear to be adaptable to nuclear power. An extensive investigation of this possibility was made in the period around 1950 in a government project NEPA (Nuclear Energy Powered Aircraft). Interest in the concept centered about the possibility of very long flights without need for refueling. The problem of shielding weight remains, and one can visualize difficulties in servicing, launching, and landing a nuclear airplane in present congested airports. It might be necessary to tow the plane to and from a remote landing strip because of the high radiation level around the reactor engine that would have to be tolerated to assure efficient operation. Of even greater concern, however, is the possibility of an airplane crash, the breakup of the light structure, and the dispersal of radioactivity over a large area.

25.2 SEA TRANSPORTATION

The success of the United States Navy's nuclear submarine fleet, the first vessel of which (1955) was the celebrated USS *Nautilus*, has demonstrated the advantage of nuclear propulsion under the sea. The submarines can operate at full power under water, have speeds that significantly exceed those of conventional vessels, and can travel for distances as far as around the world without having to be refueled. The nuclear Navy consists of some 100 submarines, developed from studies of highly enriched uranium reactors in the early 1950s, as discussed in Chapter 17. The U.S.S.R. built and has successfully operated a nuclear-powered ice breaker, the *Lenin*. Operable or in construction are three United States aircraft carriers—the *Enterprise* (with 8 reactors), the *Nimitz*, and the *Eisenhower*.

Because of security classification, it is not possible to describe the military vessels. Some appreciation of this application may be gained, however, by reference to a merchant vessel the Nuclear Ship *Savannah*,† operated for the period 1961–1971 by the Maritime Administration of the United States Department of Commerce. This combination passenger and cargo ship provided a demonstration of feasibility, but since it was the first and one-of-a-kind was uneconomical. It was 596 ft long, and accommodated 110 crew, 60 passengers, and 9400 tons of cargo. The vessel cruised at 24 knots, and was reported to be extremely maneuverable and responsive, the engine being able to go from essentially zero power to full power in 30 sec. The 22,000 shaft horsepower (16.5 MW) was provided by

†Named after the S.S. *Savannah*, the first steam-powered vessel to cross the Atlantic, in 1819.

the steam from its light water 75 MWt reactor, built by Babcock and Wilcox Co. The PWR reactor used enriched (around 4%) uranium oxide pellets in stainless steel tubes. Renewed interest in ships for the merchant marine of the United States may someday lead to the construction of a significant fleet of nuclear-powered ships.

Other countries have taken leadership in the development of nuclear ships. The West German *Otto Hahn* is in regular use for freight service; Italy has an oil tanker *E. Fermi*, and Japan has a vessel for oceanographic survey, the *Mutsu*.

25.3 NUCLEAR SPACE PROPULSION

The low fuel consumption rate makes the nuclear rocket advantageous for certain missions in space. In the project NERVA (*N*uclear *E*ngine for *R*ocket *V*ehicle *A*pplications), a cooperative effort sponsored by the United States Atomic Energy Commission and the National Aeronautic and Space Administration, successful ground tests have been made of a nuclear engine, but the system has not been used in space. The nuclear rocket engine heats hydrogen in liquid form at about $-430°F$ to a gas at temperatures around 4000°F. The high-speed hydrogen is discharged through a nozzle to provide a thrust of some 75,000 lb. The virtue of hydrogen in rocket propulsion is the low mass of expelled particles. The amount of thrust per pound of propellant (the specific impulse) is larger the lighter the element, varying inversely as the square root of the particle mass. Assuming comparable temperatures, the specific impulse for a rocket propelled by hydrogen gas would be at least twice that of one using heavier chemicals such as carbon, nitrogen, and oxygen.

Figure 25.1 shows the schematic arrangement of the NERVA engine. Liquid hydrogen from the propellant storage tank is pumped to the wall of the engine nozzle for protective cooling, and passes through the reflector. The heated hydrogen gas is next sent through the turbine that drives the pump, then returns to the core to be heated further. Finally, the gas escapes at high speed through the nozzle to give the desired thrust. The flow is controlled by the bypass control valve.

The reactor core is very compact, only about 4 ft in diameter and length. It is composed of a solid mixture of graphite and highly enriched uranium, pierced by many small (0.1-in. diameter) coolant channels. The ratio of hole area to solid area is about 0.3. Control is provided by a number of cylinders in the beryllium reflector that are part absorber and part moderator. These "drums" can be rotated for the necessary adjust-

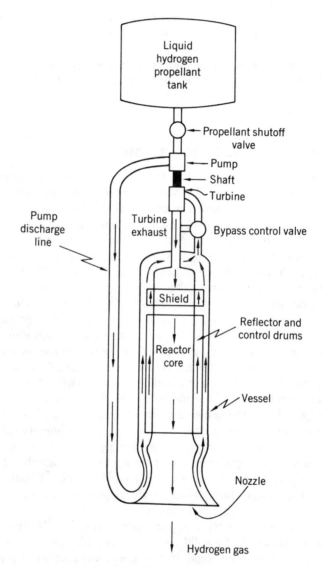

Fig. 25.1. NERVA engine.

ment of reactivity, with additional control provided by the presence of hydrogen in the reactor core. The system is designed to start up very quickly, going from about 100 MW power to 1500 MW in about 30 sec. For full power in flight, it would operate for only a couple of hours. Dr. Glenn T. Seaborg, former Chairman of the USAEC, described the NERVA reactor as "—little larger than an office desk, that will produce the 1500 megawatt power level of Hoover dam—."

The nuclear rocket would not be launched from the earth's surface since its thrust is limited and the radioactivity hazard in an aborted launch would be unacceptable. It has great promise though for powering a "shuttle service" between earth orbit and moon orbit, carrying men, equipment, and supplies to a moon base. For long voyages such as those to the planets Venus, Mars, and Jupiter, or for deep space probes, the low fuel weight of a nuclear vehicle is especially favorable.

In a manned trip to Mars, it has been proposed that the components of the interplanetary vehicle would be placed in earth orbit and assembled there. The escape from earth orbit would then be accomplished by the use of nuclear engines. Coasting for about 200 days would bring the spacecraft to Mars with braking to enter its orbit provided by the nuclear engine. Landing on the surface of Mars and return to earth would be performed by chemical rockets similar to those used in lunar exploration.

Whether or when such voyages will be made of course depends on decisions on national priorities. Although the accomplishment of manned landings on the moon was regarded as an important milestone in history, there has been growing sentiment in the United States that the solution of problems on earth needs full attention. The ultimate in sophistication for travel to distant regions of space is a fusion reactor with ions as propellant. Future centuries may see voyages of spaceships carrying colonists to new and better worlds.

25.4 SUMMARY

Propulsion by use of nuclear energy is advantageous whenever compactness and long operation without refueling is needed, because of the large energy yield from fission. Small passenger conveyances are precluded, however, because of high cost of the power supply and its initial fuel, radiation shield weight requirements, and radioactivity hazards. A large number of submarines are powered by reactors, and merchant vessels are in successful operation. A nuclear rocket engine that heats hydrogen gas to serve as propellant is available for earth–moon shuttle service and interplanetary missions.

25.5 PROBLEMS

25.1. By comparison with data on neutrons at 293°K, speed $v = 2200$ m/sec, $E = 0.0253$ eV, find the most probable speed of hydrogen molecules (H_2) at 4000°F (2477°K). Express in centimeters per second and feet per second.

25.2. (a) If the exhaust speed of H_2 propellant v_p from a nuclear rocket is 1.87† times the thermal speed (see Problem 25.1), calculate the specific impulse

$$I(\sec) = v_p (\text{ft/sec})/g (\text{ft/sec}^2),$$

where $g = 32.2$.

(b) Then, using the relation between specific impulse, propellant mass flow rate r, and thrust F,

$$F(\text{lbf}) = I(\sec)r\left(\frac{\text{lbm}}{\sec}\right),$$

find the value of r that produces 75,000 lb of thrust.

(c) If the specific heat of hydrogen gas is around 4 cal/g-°C, what is the H_2 mass flow rate at 1500 MW and coolant temperature rise from -400°F to 4000°F? How does this compare with the answer from part (b)?

25.3. Compute the ratio of the thrust produced by a nuclear rocket using molecular hydrogen as the propellant to the thrust produced by a chemical rocket that burns hydrogen and oxygen to discharge H_2O with:

(a) The same total mass flow rate.
(b) The same hydrogen mass flow rate.

Assume that the temperatures in the nuclear and chemical rocket are the same. Suggestion: Let subscripts for quantities be c for chemical, n for nuclear.

---

†The ratio is $\sqrt{\gamma/(\gamma - 1)}$, where γ is c_p/c_v, 1.40 for H_2.

26

Thermal Effects and
the Environment

The generation of electrical power by consumption of any fuel is accompanied by the release of large amounts of waste heat. As the demand for power grows, the potential effect on the environment is a matter of increasing concern. In this chapter we review the physical basis for the waste energy release, describe the available mechanisms for dealing with thermal discharges, and consider ways of turning the energy to beneficial purpose.

26.1 THERMAL EFFICIENCY

Let us examine the origin of waste heat from a plant that generates electrical power. A great deal of steam is passed through a turbine to provide the mechanical energy that drives the electrical generator, as discussed in Chapter 14. The steam leaves at a low temperature, e.g., 120°F. However, because of the latent heat of vaporization it has a large heat content, which is given up when the steam is condensed and returned as water to the steam generator. A large flow of water at a temperature near that of the surroundings must thus flow through the condenser to remove this thermal energy, which constitutes a large part of the waste heat.

For any energy conversion process the thermal efficiency e, defined as the ratio of work done to thermal energy supplied, is unfortunately limited by the temperatures at which the system operates. In accord with the second law of thermodynamics, in an ideal cycle, the highest efficiency is

$$e = 1 - \frac{T_1}{T_2},$$

where T_1 and T_2 are the lowest and highest absolute temperatures.†

To illustrate, suppose that a steam generator produces steam at 600°F and the cooling water comes from a reservoir at 70°F. The maximum efficiency for this part of the system is then

$$e = 1 - \frac{530}{1060} = 0.5.$$

The overall efficiency of the plant is lower still because of effects in piping, pumps, and turbine, reducing the efficiency for a nuclear plant to around 0.33. This means that twice as much energy is wasted as is converted to useful electrical energy. Even if we were able to use water at the freezing point 32°F, little improvement in efficiency would be achieved. A large gain would result, however, from increasing the steam temperature, say to 1000°F. Such higher steam temperatures are possible in a fossil fuel plant, giving overall efficiencies of around 0.40. The difference between 33% and 40% appears small, but for a given electrical generating capacity, say 1000 MWe, the waste heat for a typical light water nuclear plant is some 35% higher than for a modern fossil fuel plant. There are good indications that the gas-cooled thermal reactor and the liquid metal fast breeder reactor will achieve efficiencies comparable to fossil fuel plants.

26.2 HEAT REJECTION METHODS

The effect on the environment of discharging thousands of megawatts of waste heat from a generating station is intimately related to the manner in which the energy is dissipated. In order to understand the magnitude of the problem of handling condenser cooling water, let us calculate the flow needed, using simple energy considerations. Assume that the efficiency of a 1000 MWe plant is $\frac{1}{3}$, so that 2000 MW of thermal energy must be dissipated. Using the conversion factor 1 Btu/sec = 1055 W, this is 1.90×10^6 Btu/sec. Recall from Chapter 14 the energy relation $P = cM\Delta T$, where P is the power in Btu/sec, c is the specific heat (1 Btu/lb-°F for water), M is the mass flow rate in lb/sec, and ΔT is the temperature rise of

†In the Kelvin scale, these are °C + 273 or in the Rankine scale, °F + 460.

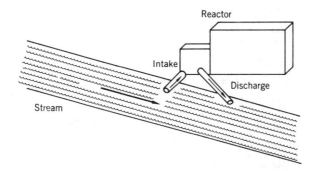

Fig. 26.1a.   Direct cooling from stream.

the condenser cooling water in °F. For a typical ΔT of 20°F, the mass flow rate is

$$M = \frac{P}{c\,\Delta T} = \frac{1.90 \times 10^6}{(1)(20)} = 9.5 \times 10^4 \text{ lb/sec,}$$

or since 1 ft^3 has mass 62.4 lb, the volume flow rate is 1520 ft^3/sec.

For reactor sites at the sea coast or next to a large inland lake, the waste heat can be readily removed by circulation and dilution. For sites near a river, the water may be extracted and returned, as sketched in Fig. 26.1a, taking advantage of natural dilution and heat transfer from the stream surface to the atmosphere to keep the temperature downstream from being excessive. From the above calculations, we see that the daily flow rate is

$$(1520 \text{ ft}^3/\text{sec})(7.5 \text{ gal/ft}^3)(8.6 \times 10^4 \text{ sec/day}) = 10^9 \text{ gal/day.}$$

A flow rate of a billion gallons per day is found only in large rivers. Even complete diversion of a small stream through the condenser would not provide the flow needed. If, instead of allowing a 20°F rise one imposes a limit on the temperature increase in a river to meet governmental regulations on water quality, say to 5°F, the stream flow would have to be at least 4 billion gallons per day.

When there is sufficient water flow in an adjacent stream but when the limit set on the increase in its temperature is too stringent, a second approach may be used. A cooling pond or lake of many acres area is constructed. As shown in Fig. 26.1b, cool water is drawn from the stream and heated water is discharged to the lake. By the time the water returns to the stream it is back to ambient temperature.

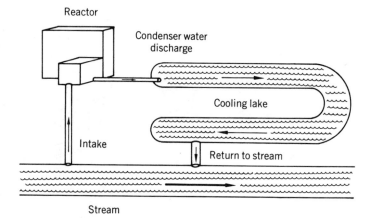

Fig. 26.1b.   Flow through cooling lake.

A third method is to isolate the cooling lake from the public waters, with the condenser cooling water drawn from and returned to the lake (Fig. 26.1c). The thermal energy deposited therein is released to the atmosphere by convection involving air currents over the surface, by evaporation, and by radiation. Makeup water from a public source is required only during periods in which rainfall does not replenish the water evaporated. We can find the upper limit on the water makeup if all heat

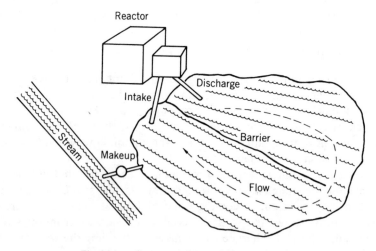

Fig. 26.1c.   Separate lake, makeup from stream.

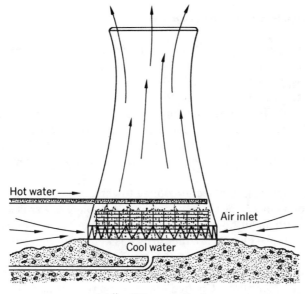

Hot water →

Air inlet

Cool water

(a) "wet" (evaporative)

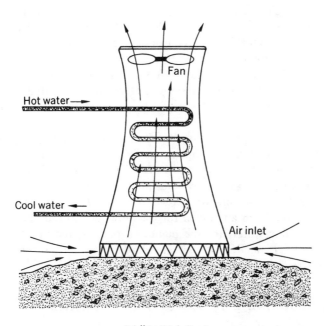

Fan

Hot water →

Cool water ←

Air inlet

(b) "dry" (air flow)

Fig. 26.2 (a and b). Cooling towers (From *Thermal Pollution and Aquatic Life* by John R. Clark. Copyright © March 1969 by Scientific American, Inc. All rights reserved.)

loss were by evaporation. The heat of vaporization of water is numerically 540 times the specific heat of water. Thus if we want to bring the temperature of a day's flow of 10^9 gal down from 20°F to 5°F, it will be necessary to evaporate (15/540) 10^9 gal, which is about 2.8% of the daily condenser flow. The higher the temperature of the discharge to a lake, the greater will be the rate of heat dissipation, and the smaller the area required. On the other hand, the effect on organisms in the water will be greater. Thus there is a competition between two desirable environmental objectives.

A fourth approach, the use of cooling towers, can be applied if land for a lake is not available or is too expensive. A cooling tower is a large heat exchanger with an air flow provided by natural convection or by blowers. In the "wet" type (Fig. 26.2a), the surface is kept saturated with moisture, and cooling is by evaporation. In the "dry" type (Fig. 26.2b), analogous to an automobile radiator, the cooling is by convection and requires greater surface area and air flow. Cooling towers for a reactor plant must be very large, of the order of 300 ft in diameter and 300 ft high. Because of the large air flow needed, winds are developed, of speeds 20 mph as far as a quarter mile from a dry tower for 2000 MWt removal. Some concern has been expressed about fog induced under certain weather conditions by the wet cooling tower, and about the large amount of makeup water required to meet losses by evaporation. It is easy to show that if the heat of vaporization is 540 cal/g the amount of water evaporated per kWhr of waste energy must be about 0.42 gal. For a 1000 MWe plant this means 20 million gallons a day.

A new concept—the offshore floating nuclear power plant—tends to eliminate the waste heat problem. A complete system is built on a platform some 400 ft on a side and towed from a shipyard to a permanent location inside a breakwater several miles out in the ocean. An underwater cable transmits electric power to a nearby city. A number of advantages have been advanced for the concept: (a) the ocean provides unlimited cooling water, (b) the plant can be near the point of greatest electrical power need, and (c) standardization and mass-production techniques reduce delays and costs.

26.3 BIOLOGICAL EFFECTS OF HEATED WATER

Our complex ecological system is sensitive to changes in many environmental conditions, including temperature. However, aside from stating the fact that most organisms cannot exist at elevated temperature, few generalizations can be made, and in order to assess the impact of heated water, it is necessary to examine the existing balance of species of plants and animals at each location. Individual species vary in their ability to thrive in water at abnormal temperatures. For example, salmon and trout can live only within a narrow range of temperatures while oysters and barnacles can stand wide variations.

An increase in temperature causes an increase in metabolism of animals that live in the water, doubling roughly with each 10°C increase. The need for oxygen goes up in proportion, but there is a tendency for oxygen content in water to decrease as the temperature is raised. In addition, the organism's ability to assimilate oxygen is reduced.

Laboratory tests provide information on upper lethal temperature limits. These range from 106°F for tropical fish to 77°F for salmon and trout. Some of the recommended maximum temperatures for well-being and growth are 93°F for catfish and bass, but 84°F for pike and perch; those for spawning and egg development are 75°F for bass, 55°F for salmon, and 48°F for trout.

The rates of growth of plants such as algae depend on temperature, with those causing disagreeable changes in odor or taste of water being favored by higher temperature. Also, there is an increase in the toxicity of poisons in the water, e.g., chlorine as used to clean condenser tubes, or pesticides such as DDT.

Tests on confined specimens do not reveal information on the ability of a mobile animal to avoid an adverse environment. However, if the region of excessive temperature is too large, the normal path of migration may be seriously disturbed. An increase in temperature may trigger spawning or migration at an earlier time than normal, which can affect the chances of survival.

On discharge from a power plant, heated water mixes with and is diluted by the water of the stream, lake, or ocean, and a negligible temperature rise is observed at large distances. It is recognized that there must be a mixing zone, but it is recommended that the zone be made as small as possible and not block passage for migration. It should be noted, however, that increases in temperature may be desirable, especially in colder climates. Conventional power plants have long observed the abundance of fish at the condenser discharge, presumably because of the increase in plant food supply at that point.

A legal basis has been established in the United States for the control of thermal pollution through various Acts of Congress, with responsibility for enforcement placed on the states. Recommended limits, for waters outside mixing zones, have been provided on maximum temperature and on the increase resulting from the installation of a power plant. Typical limits range from 1.5°F to 5°F, dependent on the season of the year and on whether the water is fresh or marine.

Much more information is needed on the complex effects of temperature on the ecology, in order to ascertain the extent of disturbance of natural conditions and the economic impact. Unfortunately, effects on the complex system are difficult to obtain without creating a significant disturbance. The imposition of an arbitrary small limit on temperature may be unnecessarily strict, unless it is assumed that our environment must not be changed in any way. Finally, political decisions must be made as to the importance of changes in the plant and animal population distribution, in comparison with power needs and the cost of preventing adverse effects.

26.4 BENEFICIAL USES OF WASTE HEAT

Much consideration has been given to the utilization of the waste heat for beneficial purposes, but relatively few large-scale applications have been initiated as yet. Some of the concepts that have been advanced are as follows:

(a) Use of the energy for home, office, and factory heating, by means of a central community system. In order to use the energy effectively, it would be necessary to plan and construct the heating and air-conditioning systems for the whole city in conjunction with power plant development. Since urban needs vary with climate and season, optimization is difficult. However, the concept has been successfully demonstrated in a small nuclear plant near Stockholm, Sweden, producing 10 MW of electricity and 80 MW of hot-water power.

(b) Enhancement of agricultural production. Warm water can be used to heat the ground and thus to stimulate growth or to increase the number of harvests of a crop per year, or to permit raising varieties that are not normally possible because of climate. Large-scale controlled production of food fish such as catfish is feasible with a heated water supply. The demand for heat varies with the season, of course, requiring alternate channels for dissipation.

(c) Desalination of salt water or brackish water. By use of the steam discharged from a turbine, desalination to produce water fit for drinking and agricultural use is feasible, with some modifications and compromises in design. The reactor system might be for the sole purpose of desalination or be a dual purpose system that optimizes for both electricity and process heat.

The concept of an agro-industrial complex built around nuclear power plants is very attractive. One can visualize the use of the electricity for manufacturing and for domestic use in the community; the discharge steam for desalination of sea water to provide water for human and animal consumption, industrial use, and for irrigation; the application of heated water for temperature control in homes and other buildings; the use of power for production of fertilizer and for pumping irrigation water; the processing of minerals extracted from the sea water; and so on.

26.5 SUMMARY

Large amounts of waste heat are discharged by electrical power plants because of inherent limits on efficiency. For typical nuclear systems, a billion gallons of water per day must be passed through the steam condenser to ensure that the temperature rise in the environment is small. The water is taken from large rivers or artificial lakes, or cooling towers are employed. Heated water can have adverse effects on plant and animal life with a great variation in sensitivity among species. Potential beneficial uses of waste thermal energy include space heating, enhancement of agricultural production, and desalination of sea water.

26.6 PROBLEMS

26.1. The thermal efficiencies of a PWR converter reactor and a fast breeder reactor are 0.33 and 0.40, respectively. What are the amounts of waste heat for a 900 MWe reactor? What percentage improvement is achieved by going to the breeder?

26.2. As sketched, water is drawn from a cooling pond and returned at a temperature 25°F higher, in order to extract 1500 MW of waste heat. The heat is

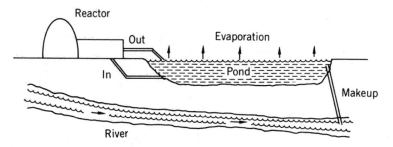

dissipated by water evaporation from the pond with an absorption of 1000 Btu/lb. How many pounds per second of makeup water must be supplied from an adjacent river? What percentage is this of the circulating flow to the condenser?

26.3. As a rough rule of thumb, it takes 1–2 acres of cooling lake per megawatt of installed electrical capacity. If one conservatively uses the latter figure, what is the area for a 1000-MWe plant? Assuming 35% efficiency, how much energy in Btu is dissipated per square foot per hour from the water? Note: 1 acre = 4.356×10^4 ft^2.

27

Energy and Resources

The world has awakened recently to the realization that we are facing three related major problems—environmental pollution, the depletion of natural resources, and a population explosion. Fundamental to these is the potential shortage of energy, which is basic to meeting all of man's physical needs—clothing, shelter, transportation, convenience, and recreation, to name a few. In this chapter we shall look at the trends in energy consumption, the possible alternative sources, and the role of fission and fusion processes in man's future.

27.1 TRENDS IN ENERGY CONSUMPTION

The use of energy by a country is closely correlated with its degree of technological development and industrialization, which are in turn related to the people's standard of living. Figure 27.1 shows the past and predicted future trends in energy consumption in the United States. We see that wood was the main source a hundred years ago. The growth of coal usage in the latter part of the 1800s and early 1900s is associated with rapidly increasing industrial development. There followed an enormous expansion in the consumption of natural gas and oil for heating, electrical generation, and especially transportation. The total consumption of energy in the United States is expected to double during the last twenty-five years of this century, with nuclear energy playing a significant role. The main reason for the unusual growth is the increasing production of goods and services that have been demanded or accepted for improved living conditions.

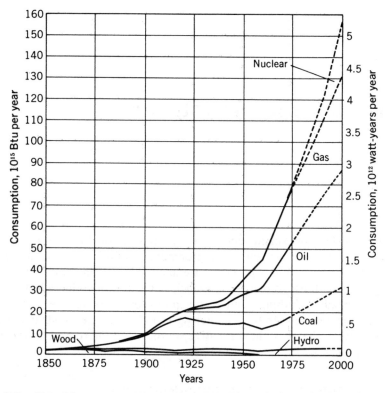

Fig. 27.1. United States energy consumption. (From *Energy and Power* by Chauncey Starr. Copyright © September 1971 by Scientific American, Inc. All rights reserved.)

One obvious example is the use of the automobile, which has become an essential part of modern existence. The widespread adoption of air conditioning for comfort, efficiency, and health has had the effect of reversing the season of peak demand for electricity from winter to summer.

On a world basis, the average level of industrial development and accompanying standard of living is far behind that of the United States and technologically advanced nations of Europe, and the developing countries are seeking to improve their conditions as rapidly as possible. However, the gap between developing and technologically advanced countries tends to widen, as population growth due to improved health exceeds the rate of industrialization. Dr. Chauncey Starr comments:† "If

†In the lead article in the September, 1971, issue of *Scientific American*, which is devoted entirely to the subject of energy.

the underdeveloped parts of the world were conceivably able to reach by the year 2000 the standard of living of Americans today, the world-wide level of energy consumption would be roughly 10 times the present figure."

Energy usage in the United States is as listed below:

Use	Percent
Residential ⎤ Commercial ⎦	22
Industrial	37
Transportation	25
Losses	16

More than half of the energy consumed in the home and business sections is for space heating; the rest is mainly for air conditioning, lighting, cooking, and appliances. Of the transportation part, autos, trucks, and buses use three-fourths of the energy. The main industrial uses are for production of metals and oil.

Of the present energy consumed in the United States, 25% goes to produce electricity; before the end of the century it is expected to increase to 40%, because of electricity's versatility, cleanliness, and ease of distribution. If we were to develop and adopt electric automobiles and high-speed electric trains for public transportation, as discussed in Chapter 25, the figure would go even higher. No estimate has yet been made of the potential electrical demand for control of environmental emissions, but it is well known that almost every device considered uses electricity.

The bulk of materials now used for energy are classified as fossil fuels—of plant and animal origin, deposited in the earth over millions of years in the form of coal, oil, and natural gas. They are classed chemically as hydrocarbons, involving the elements carbon, hydrogen, oxygen, and nitrogen. The fossil fuels are extremely useful raw materials because of the conveniently stored chemical energy, but if they are burned for fuel, the products are released and lost. Even if we disregard the pollution in the form of oxides of nitrogen, carbon, and sulfur that results from the burning, there is a staggering waste of natural resources that will never be available again. It is preferable to use the fuels for the production of more permanent and valuable products, and then to recycle them when no

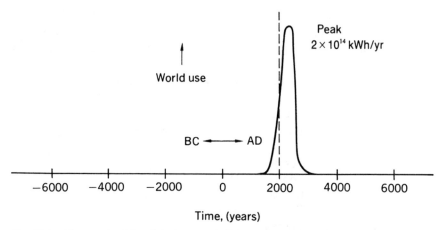

Fig. 27.2. The epoch of fossil fuels. (Adapted from *Energy Resources: A Report to The Committee on Natural Resources,* Publication 1000-D, M. King Hubbert, National Academy of Sciences—National Research Council, Washington, D.C., 1962.)

longer useful. Such practices are preferable to burning of wastes, which in turn is better than continued dumping of wastes.

Modern man is using the resources at such a rate that their exhaustion within a very few centuries is likely. This may seem to be a long time in comparison with a generation or even a lifetime, but in the span of man's history, it is a very brief epoch. A dramatic statement of this problem is provided by the graph† in Fig. 27.2.

27.2 ALTERNATIVE SOURCES OF ENERGY

If the world is to solve the long-range energy problem, it must make use of all available practical sources. Table 27.1 shows the main possibilities, classified as fossil, physical, fission, and fusion. Let us review the situation on each of these.

Natural Gas

As a byproduct of petroleum production, natural gas has been abundant until recent years, when the rate of discovery of new wells has decreased. It is a popular source of energy because of its convenience for use and

† Adapted from that of Dr. M. King Hubbert in his article, "Energy Resources," in a most readable and significant book *Resources and Man.*

Table 27.1 Energy Sources.

FOSSIL	Animal and plant origin
Wood	
Coal	
Oil	
Gas	
PHYSICAL	Solar and terrestrial
Hydro	
Wind	
Tidal	
Geothermal	
Solar	
FISSION	Nuclear, heavy elements
Burner	
Converter	
Converter with Pu recycle	
Breeder	
FUSION	Nuclear, light elements
Deuterium–tritium	
Deuterium–deuterium	

cleanliness in burning, and because of controlled low prices. Although there is a possibility of stimulation of gas production by underground explosives as described in Chapter 24, it is generally believed that gas will be the first fossil fuel to be in short supply.

Some thought has been given to the use of hydrogen gas as a fuel. Evolved by the electrolysis of water, its burning is pollution-free, and it is as readily piped and as safe as natural gas. Its use does not eliminate the need for energy to effect the dissociation of water, of course.

Oil

The discovery of new reserves in Alaska and the decision to install a pipeline to bring the oil out will increase the supply of oil to the United States, but is not expected to meet the long-term demands for oil. Expanded off-shore drilling for new oil will also expand reserves, with accompanying increased environmental concerns, of course. Large deposits of oil-bearing shale rock are available in the western part of the United States, but the cost of extraction of oil is high, and the increase in volume resulting from processing poses a serious disposal problem. The importation of large amounts of oil from the Middle East has many implications related to political uncertainty and international balance of payments, along with problems of adequate shipping, storage, and refining capacity.

Coal

The reserves of coal are variously estimated to be adequate for 100–400 yr, but the cost of extraction continues to increase, in part because of the introduction of badly needed mine safety measures. The effect of strip mining on the land is objectionable, and the problem of control of all gaseous emissions has yet to be solved. The development of an economic method of gasification of coal would permit the elimination of undesirable byproducts such as sulfur prior to burning. The alternative or parallel development of coal liquefaction would allow coal to be extracted from deep underground deposits. The long-range fossil fuel energy problem would still be with us, however.

Fuel cells, which convert chemical energy directly to electrical energy, have the advantage of high efficiency, reduced emission of pollutants, and ability to provide local sources of energy. Units under investigation are yet too small to make a major contribution to resource conservation.

The burning of fossil fuels to produce heat that is transformed into mechanical and then electrical energy is an inherently inefficient process, as discussed in Chapter 26. Improvements in efficiency of conventional steam plants are possible with "topping" cycles, which carry a working fluid to very high temperature. Liquid metals and gases are good candidates because of their thermal properties. Another approach to improved efficiency is the use of magnetohydrodynamic generation, in which a stream of high-temperature ionized gas replaces the moving conductor in an electrical generator, yielding direct current at high potential. As a topping cycle for conventional steam plants, an increase in efficiency from 40% to 50% may be feasible, but there are many technical and economic problems yet to be solved.

Hydroelectric power is available when a stream can be dammed to provide a large reservoir, permitting falling water to turn a hydraulic turbine coupled to an electric generator. It is generally agreed that most of the good sites for hydropower are already in use, and that this source will not meet the total need. The pumped storage technique, in which the turbines are used to pump water up to a reservoir, helps accommodate load demands, but generates no net energy.

Geothermal power, coming from the heat in the earth's crust, is available at a few sites, notably Italy and California, where the geological formation favors the natural release of steam. The environmental aspects of this source of energy such as saline waste water disposal and gas emissions, are not fully known, and the number of readily accessible sites is limited. There is a possibility, as yet unexplored, that the abundant heat of the earth could be tapped at many locations by drilling deep enough.

Tidal power is a less conventional method, in which water from the sea enters and leaves a restricted channel periodically, with each stream turning a turbine. Only one full-scale plant is in existence, on the coast of France, and there are a few other promising sites where tides and terrain are suitable. For the long-range energy need, however, this approach will not be adequate.

Wind power was at one time an important source of energy on farms, and has been considered recently for use in regions where strong winds prevail. An enormous number of windmills would be required, accompanied by electric storage facilities. The aesthetic aspect of windmill power must be considered, over and beyond the technical and economic factors.

Solar power is often mentioned as the logical and proper alternative, and indeed, the amount of radiant energy that strikes the earth's surface is far more than is needed. There is good evidence that individual homes can be built to take advantage of the energy of sunlight for temperature control. The incentives of fuel shortages and economic factors could result in their adoption.

For the generation of electricity, however, there are serious problems to be solved. To collect and concentrate the energy by reflectors and converters of present efficiency is the major difficulty. An area of collector of about 16 mi^2—4 miles on a side—would be needed to supply electricity to a city of a couple hundred thousand population, with efficiencies of say 20%. Obvious problems come to mind, related to unfavorable weather conditions, energy storage requirements, and the use of the shadowed space. The desert is a logical place to locate solar devices because of open space, isolation, and the frequency of sunny days, but transmission costs for electrical power to urban centers would be excessive. We can hope to develop converters from heat energy to electrical energy that are more efficient than 20%, and perhaps exploit superconductivity of metals at very low temperatures to reduce transmission costs, but it is clear that there remain many technological problems in this area.

A recently advanced concept for harnessing solar energy is intriguing. It is proposed to place collectors of the sun's rays in synchronous orbit, 22,300 miles out in space, with the energy that is absorbed by solar cells transmitted to the earth by microwaves that would easily penetrate the atmosphere. The radiation would then be collected, rectified, and transmitted through superconducting lines. The feasibility of the idea is uncertain at this time.

In our discussion of energy sources, a tacit assumption has been made that national growth patterns will continue indefinitely, which is impossible both practically and mathematically. It is likely that the per capita consumption, in the United States at least, will level off as most of the needs are met. There are indications also that the rate of growth of population is declining in the United States, with a possible plateau of some 300–400 million people. These trends can be significantly affected by public attitudes. For example, the recognition of the seriousness of the energy problem could stimulate strong national efforts to minimize the waste of energy by a variety of techniques. Included are the adoption of improved building insulation, the extension of useful life of manufactured goods and equipment, and restraint in the use of gasoline and electricity, either on a voluntary basis, or through mechanisms of price control or rationing. The wisdom of conservation is undeniable both from the standpoint of resource conservation and environmental protection. Advanced countries, and especially the United States, could readily adopt such conservation practices without serious inconvenience, but there remains the problem of bringing the rest of the several billions of people in the world up to an acceptable standard of living, with the accompanying drain on energy resources.

27.3 THE ROLE OF NUCLEAR ENERGY

Several conclusions can be derived from the foregoing discussion: that efforts to eliminate the extravagant waste of energy are needed; that research and development on new sources and ways of increasing efficiency are in order; and that sources alternative to fossil fuels must be fully developed and utilized wherever appropriate. We are inevitably led to give serious consideration to nuclear energy, in spite of potential

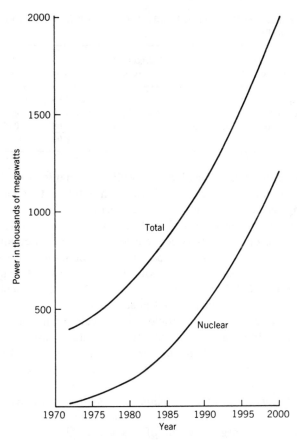

Fig. 27.3.   United States electric power.

hazard. Predictions of the amounts of electrical generation needed and the part to be contributed by nuclear reactors in the next few decades are shown in Fig. 27.3. By the year 2000, about half of the electrical supply is expected to come from nuclear sources.

There are many factors that will determine the accuracy of this prediction. First is the question of public acceptance of nuclear plants. If those who regard the potential hazard of accidental radioactivity release as a sufficient deterrent to continued adoption of nuclear reactors are able to convince the public of their view, the trend would clearly go to zero. If demands that additional protective equipment be developed and installed result in an excessive construction and operating cost of nuclear plants,

Table 27.2. AEC—Estimated Costs (1981) for 1000 MWe Power Plants in mills/kWh.

Component	Light water reactor	Coal	Oil
Capital cost	11.70	10.90	8.00
Fuel cost	2.50	5.50	24.60
Operation and maintenance	1.00	1.60	0.80
	15.20	18.00	33.40

there will be a tendency for electrical utility companies to shift back to coal-fired plants, and the relative distribution between fossil and nuclear would change greatly. If, however, concerns increase about the emissions of fossil plants and about fossil fuel shortages with sharply increased costs, or if discrepancies between energy supply and demand become excessive, the growth of nuclear power might be even more rapid than anticipated. Comparison of costs of electrical power from the main sources are given in Table 27.2. It is noted that the cost of construction of a nuclear plant is high but the fuel cost is low.

Let us examine the various nuclear fuel cycles that are presently or potentially available. Rates of consumption of natural resources and costs are both strongly dependent on the system employed. In order of effectiveness of conversion of mass-energy into electrical energy, the cycles are as follows:

(a) The *burner* utilizes only uranium that is highly enriched (e.g., 93%) in the isotope U-235, with separation achieved by gaseous diffusion plants. There is a large amount of discarded U-238 in the depleted uranium, and the consumption of natural uranium is excessive. Figure 27.4a shows the fuel cycle schematically. Such reactors are likely to be used only for special purposes such as propulsion of submarines, surface vessels, and space vehicles.

(b) The *converter* makes use of slightly enriched uranium (e.g., 2–3%), with a smaller amount of depleted uranium discharged in the isotope separation process than for the burner, and with the generation of Pu-239 from the U-238 in the reactor fuel. Some power is produced by the plutonium, and some is discharged and stored (see Fig. 27.4b). Most of the reactors being installed in the 1970s are of this light-water-moderated type. There are two competitors, however, as noted in Chapter 13. One is the natural uranium fueled heavy water reactor, which does not require the enrichment process, the other is the high-temperature gas-cooled

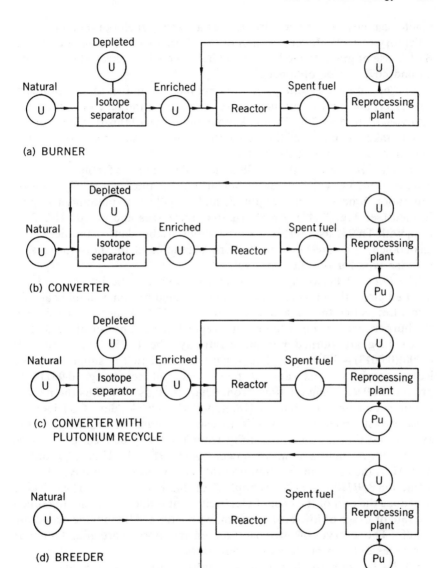

Fig. 27.4. Fission fuel cycles.

reactor employing enriched uranium as a starting fuel, but building up an inventory of U-233 by conversion of the alternate fertile element Th-232. By the use of graphite moderator and helium coolant, thermal efficiencies around 40% can be obtained.

(c) The *converter with plutonium recycle* is similar to the ordinary converter, but the Pu-239 that is discharged from the reactor is fed back into the reactor (see Fig. 27.4c). A movement toward this recycle is under way to take advantage of the fissile plutonium, burning it as a fuel rather than having to pay storage costs.

(d) The *breeder* uses the U-238 in natural uranium as fertile material to produce Pu-239, which is the principal fuel for the reactor. The isotope separation process is not required, and recycling of plutonium is performed (see Fig. 27.4d). An alternative cycle uses and breeds U-233 as fuel, with Th-232 as a fertile material about as abundant as uranium. The gas-cooled, graphite-moderated thermal reactor is the most promising breeder type for the thorium cycle.

The contrast between the least efficient system, the burner, and the most efficient, the breeder, can be demonstrated by some simple calculations. Let us find the consumption rate of U-235 (in kilograms per day) in the burner that is needed to obtain a typical thermal power of 3000 MW. Since 1.3 g are burned per megawatt-day, the U-235 requirement is $(3000)(0.0013) = 3.9$ kg/day. This highly enriched fuel is supplied by the isotope separator. Its ratio of feed at weight percentage 0.00711 to product at 0.93, using relations from Chapter 10, is 182. Thus the natural uranium consumption rate is $(182)(3.9/0.93) = 762$ kg/day. Consider instead a fast breeder that is able to use all natural uranium fed to the system. Its power comes primarily from Pu-239, produced as U-238 is consumed. A small amount of power comes from the U-235. To obtain 3000 MW, the uranium consumption rate is now only 3.9 kg/day, which is a fraction $3.9/710 = 0.0055$ or about $\frac{1}{2}\%$ of that in the burner. The effect is to reduce the part of the cost of electricity that is due to fuel and to reduce the rate of usage of low-cost uranium, with large accompanying economic benefit. The reserves of low-cost uranium and thorium are quite limited if one relies only on burners and converters.

(e) *D–T fusion* would use deuterium separated from ordinary hydrogen in water and tritium produced by neutron bombardment of lithium. The supply of the latter is the ultimate limitation on long-term use of the D–T reaction. Figure 27.5 shows a hypothetical fusion cycle.

(f) *D–D fusion* would use deuterium only, and thus require higher plasma temperatures. This is the ultimate energy source other than solar energy.

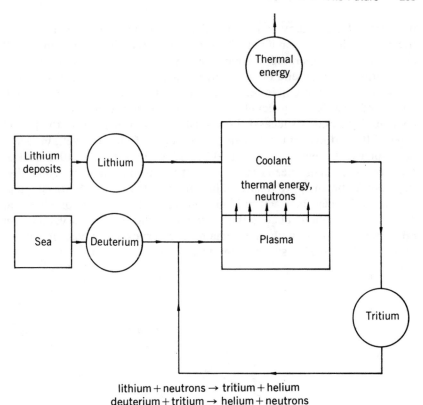

lithium + neutrons → tritium + helium
deuterium + tritium → helium + neutrons

Fig. 27.5. Fusion fuel cycle.

27.4 THE FUTURE

Let us try to visualize the pattern of development and application of these systems over the next hundred years or so, realizing that long-range predictions may be very far off the mark. Nuclear plants of the converter type, depending on slightly enriched uranium and producing certain amounts of plutonium, will probably continue to be built and will contribute an increasing fraction of the electrical supply. Considerable research and development will be devoted to the assurance of safety. There will be gradual adoption of the plutonium recycle in view of two factors: (a) the mounting stockpile of that element along with the cost of storage, and (b) the ability to reduce the U-235 enrichment in the uranium used as fuel. Current research indicates that the breeder reactor will be demonstrated as practicable and commercially advantageous during the 1980s, and a number of new breeders will subsequently be put into operation. Since many converter plants built prior to the advent of breeders will have remaining lifetimes of up to 30 years, they will continue to be operated, helping supply plutonium for new breeder installations. By the end of the century, there will be a mixture of about the same number of converters and breeders in operation. In the interim, continued research and development on the fusion process has a good chance of yielding a successful demonstration power plant in the period 2000–2010, and plant installations would begin. If all goes well, breeders could phase out over a period of 50 years; and by the year 2100, most of the electrical energy needs might be met by fusion reactors. Hopefully, there would be an approximate stabilization of the United States population at no higher than 500 million. The fossil fuel usage for electrical power could reach its peak in the early part of the next century, and drop off nearly to zero, with the remaining reserves employed for other purposes. Figure 27.6 suggests the trends just described.

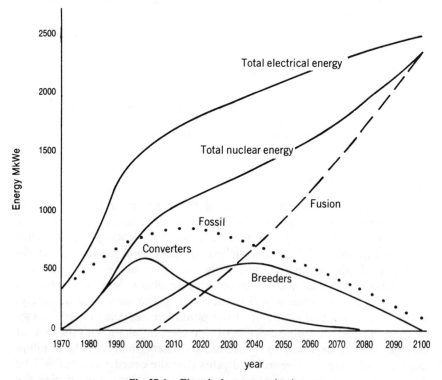

Fig. 27.6. Electrical energy projections.

A new era of existence for man can be imagined. An essentially unlimited fuel supply would be available from the deuterium of the oceans. The energy would be used to manufacture all of man's physical requirements, to recycle all waste materials for reuse, and to make possible new levels of technological and cultural accomplishment. If all needs are provided by abundant and inexpensive energy, harmony among nations can be visualized.

27.5 SUMMARY

A very rapid growth in energy consumption has occurred in the twentieth century, especially in the United States. The principal source—fossil fuels—is likely to be in short supply immediately and exhausted in a few hundred years, requiring that alternative sources be found. Hydroelectric, wind, tidal, geothermal, and solar power require much technological development. Conservation practices are desirable to avoid energy waste. Nuclear energy fuel cycles include the burner, converter, and breeder, the latter providing a great reduction in uranium consumption. Fusion reactors if successfully developed would allow us to reduce fossil fuel usage to zero within a century and provide man's energy needs indefinitely.

27.6 PROBLEMS

27.1. Find the necessary collector area in square miles for solar radiation if the efficiency of conversion from thermal energy to electrical energy is 20% and the United States is provided 10^9 kW of electrical power, as is expected to be needed before 1990. Assume that the incident radiation is available 4000 hr/yr at 300 W/m². (Note: 1 mi² $= 2.59 \times 10^6$ m².) Compare the result with the area of the State of Arizona, 113,909 mi².

27.2. What is the consumption rate of natural uranium for the converter system (Fig. 27.4b), if the ratio of feed to product in the isotope separator for product at 3% U-235 is 5.5 and spent fuel is not recycled? Compare the answer with that for the burner and the breeder.

27.3. If 1.3 g of U-235 are consumed per megawatt-day of thermal energy, what is the corresponding value for Pu-239? Note cross sections: U-235 $\sigma_f = 582$, $\sigma_a = 681$; Pu-239 $\sigma_f = 742$, $\sigma_a = 1011$.

27.4. In a thermal converter system with plutonium recycle (see Fig. 27.4c), the ratio of consumption rates of Pu-239 and U-235 in the 3000-MWt reactor is $\frac{2}{3}$. Assume that the makeup fuel supplied to the reactor is uranium at 1% U-235 coming entirely from the isotope separator. Let the feed to the separator be 0.711% U-235, and the waste from the separator 0.2% U-235.

(a) Find the consumption rate of U-235 in the reactor (kg/day) (which must also be the makeup rate in the closed system).

(b) Find the feed rate of natural uranium to the separator (kg/day). Note: Per megawatt-day, the weights of fuel consumed are U-235 1.3 g, Pu-239 1.5 g.

27.5. If the total United States electrical energy consumption by the end of the twenty-first century is 2500 million kilowatts, and 95% is provided by fusion using the D–D reaction at a thermal efficiency of 0.5, how much deuterium will be required per year, and what will the fuel cost be (see Problem 16.4)?

Appendix   Selected References

Chapter(s)	Reference
General	*Understanding the Atom Series* (or *World of the Atom Series*), United States Atomic Energy Commission, Division of Technical Information (or Office of Information Services), P.O. Box 62, Oak Ridge, Tennessee, 37830. Titles in this set, cited elsewhere also as "UAS" are:

Accelerators
Animals in Atomic Research
Atomic Fuel
Atomic Power Safety
Atoms at the Science Fair
Atoms in Agriculture
Atoms, Nature, and Man
Books on Atomic Energy for
  Adults and Children
Careers in Atomic Energy
Computers
Controlled Nuclear Fusion
Cryogenics, The Uncommon
  Cold
Direct Conversion of Energy
Fallout from Nuclear Tests
Food Preservation by
  Irradiation
Genetic Effects of Radiation
Index to the UAS Series
Lasers
Microstructure of Matter
Neutron Activation Analysis

Nondestructive Testing
Nuclear Clocks
Nuclear Energy for Desalting
Nuclear Power and Merchant
  Shipping
Nuclear Power Plants
Nuclear Propulsion for Space
Nuclear Reactors
Nuclear Terms, A Brief Glossary
Our Atomic World
Plowshare
Plutonium
Power from Radioisotopes
Power Reactors in Small Packages
Radioactive Wastes
Radioisotopes and Life Processes
Radioisotopes in Industry
Radioisotopes in Medicine
Rare Earths
Research Reactors
SNAP, Nuclear Space Reactors
Sources of Nuclear Fuel
Space Radiation

Chapter(s)   *Reference*

General

Spectroscopy	The First Reactor
Synthetic Transuranium Elements	The Natural Radiation
The Atom and the Ocean	Environment
The Chemistry of the Noble Gases	Whole Body Counters
The Elusive Neutrino	Your Body and Radiation

A Bibliography of Basic Books on Atomic Energy, 1971. UAS

Environmental Aspects of Nuclear Power Stations. International Atomic Energy Agency, Vienna, 1971. Proceedings of a symposium, New York, August 10–14, 1970, organized by IAEA in cooperation with USAEC. Reference cited as *New York 1970.*

H. Etherington (Ed.) *Nuclear Engineering Handbook.* McGraw-Hill, New York, 1958.

S. Glasstone. *Sourcebook on Atomic Energy.* Van Nostrand, New York, 1950.

Nuclear News, monthly publication of the American Nuclear Society, Hinsdale, Ill.

Nuclear Terms, A Brief Glossary, 1967. UAS

Peaceful Uses of Atomic Energy. Proceedings of the fourth international conference, Geneva, September 6–16, 1971. United Nations, New York and The International Atomic Energy Agency, Vienna, 1972–73. 15 volumes. Reference cited as *Geneva 1971.*

K. B. Pedersen, *et al. Applied Nuclear Power Engineering for Practicing Engineers.* Cahners Books, Boston, 1972.

Reactor Safety, quarterly publication of the United States Atomic Energy Commission. Office of Information Services.

Reactor Technology, quarterly publication of the United States Atomic Energy Commission. Office of Information Services. Formerly *Power Reactor Technology,* published 1957–1973.

G. T. Seaborg and W. R. Corliss. *Man and Atom.* E. P. Dutton, New York, 1971.

1   Raymond L. Murray and Grover C. Cobb. *Physics Concepts and Consequences.* Prentice-Hall, Englewood Cliffs, N.J., 1970.

1–8   R. E. Lapp and L. H. Andrews. *Nuclear Radiation Physics.* Prentice-Hall, Englewood Cliffs, N.J., 1965.

2   Clifford E. Swartz. *Microstructure of Matter,* 1965. UAS

3   Jacob Kastner. *Nature's Invisible Rays,* 1970. UAS

5   *Neutron Cross Sections,* BNL 325 Second Edition, Supplement No. 2, Volumes I, IIA, B, and C, III. 1964–66. Available from National Technical Information Service, Springfield, Va.

5   S. F. Mughabghab and D. I. Garber. *Neutron Cross Sections Volume 1, Resonance Parameters,* BNL 325 Third Edition, Volume 1, Brookhaven National Laboratory, Upton, N.Y., June 1973.

5, 12, 13   P. J. Grant. *Elementary Reactor Physics.* Pergamon Press, Oxford, 1966.

8   Harold Grad. "Plasmas," *Physics Today,* December 1969, pp. 34–44.

9   M. Stanley Livingston and John P. Blewett. *Particle Accelerators.* McGraw-Hill, New York, 1962.

Chapter(s)	*Reference*
	R. R. Wilson, "The Batavia Accelerator," *Scientific American*, February 1974, pp. 72–83.
10	Vincent V. Abajian and Alan M. Fishman. "Supplying enriched uranium," *Physics Today*, August 1973, pp. 23–29.
	L. O. Love. "Electromagnetic separation of isotopes at Oak Ridge," *Science*, October 26, 1973, pp. 343–352.
	William D. Metz. "Uranium Enrichment: U.S. 'One Ups' European Centrifuge Effort," *Science*, March 29, 1974, pp. 1270–1272.
13	Andrew W. Kramer. *Understanding the Nuclear Reactor.* Power Engineering Magazine, Technical Publishing Co., Barrington, Ill., 1970.
14	Summers, Claude M. "The conversion of energy," *Scientific American*, September 1971, p. 148.
15	Walter Mitchell, III and Stanley E. Turner. *Breeder Reactors*, 1971. UAS
	Alfred M. Perry and Alvin M. Weinberg. "Thermal breeder reactors," *Annual Review of Nuclear Science*, Vol. 22, 1972, pp. 317–354.
	Glenn T. Seaborg and Justin L. Bloom. "Fast breeder reactors," *Scientific American*, November 1970, p. 13.
15, 27	Arthur L. Singleton, Jr. *Sources of Nuclear Fuel*, 1968. UAS
16	Bruno Coppi and Jan Rem. "The Tokomak approach in fusion research," *Scientific American*, July 1972, p. 65.
	William C. Grough and Bernard J. Eastlund. "The prospects of fusion power," *Scientific American*, February 1971, p. 50.
	Moshe J. Lubin and Arthur P. Fraas. "Fusion by laser," *Scientific American*, June 1971, p. 21.
	John Nuckolls, John Emmett, and Lowell Wood. "Laser-induced thermonuclear fusion," *Physics Today*, August 1973, pp. 46–53.
	Richard F. Post. "Prospects for fusion power," *Physics Today*, April 1973, pp. 30–39.
17	H. D. Smyth. "Atomic energy for military purposes," *Reviews of Modern Physics*, Vol. 17, No. 4, pp. 351–471. The first unclassified account of the nuclear effort of World War II. Readily understood technical information and administrative history of the Manhattan Project. See also the book version, listed below.
	H. D. Smyth. *Atomic Energy for Military Purposes.* Princeton University Press, Princeton, N.J., 1945.
18	Isaac Asimov and Theodosius Dobzhansky. *The Genetic Effects of Radiation*, 1966. UAS
	"The effects on populations of exposure to low levels of ionizing radiation" (the BEIR Report). National Academy of Sciences—National Research Council, Washington, D.C., November 1972.
	Daniel S. Grosch. *Biological Effects of Radiations.* Blaisdell, New York, 1965.
	Walter E. Kisieleski and Renato Baserga. *Radioisotopes and Life Processes*, 1967. UAS
	Joseph A. Lieberman. "Ionizing-radiation standards for population exposure," *Physics Today*, November 1971, pp. 32–38.
	Ruth Moore. *Evolution.* Life Nature Library, Time, Inc., New York, 1962.

Chapter(s) | *Reference*

John Pfeiffer. *The Cell.* Life Science Library, Time, Inc., New York, 1964.

"Radiation effects on man," *Nucleonics 21,* 1963, pp. 45–60.

W. L. Russell. "The genetic effects of radiation," *Geneva 1971,* Vol. 13, p. 487.

18, 19   *Radiological Health Handbook.* U.S. Department of Health Education and Welfare, Rockville, Md., 1970.

19   K. Z. Morgan and E. G. Struxness. "Criteria for the control of radioactive effluents," *New York 1970,* p. 211.

T. Rockwell, III (Ed.) *Reactor Shielding Design Manual.* McGraw-Hill, New York, 1956.

20   John F. Hogerton. *Atomic Power Safety,* 1964. UAS

Charles K. Leeper. "How safe are reactor emergency cooling systems?," *Physics Today,* August 1973, pp. 30–35.

Nuclear Power and the Environment, 1969. UAS

The Safety of Nuclear Power Reactors (Light Water Cooled) and Related Facilites. United States Atomic Energy Commission, Washington, D.C., July 1973. Report WASH-1250.

S. E. Rippon. "Light water reactor safety," A review of WASH-1250, *Nuclear Engineering International,* January 1974, pp. 25–30.

L. D. Smith. "Evolution of opposition to the peaceful uses of nuclear energy," *Nuclear Engineering International,* June 1972, pp. 461–468.

20, 21   R. Philip Hammond. "Nuclear power risks," *American Scientist,* March–April 1974, pp. 155–160.

D. Bruce Turner. *Workbook of Atmospheric Dispersion Estimates,* Public Health Service, Cincinnati, Ohio, 1970.

21   W. J. Bair and R. C. Thompson. "Plutonium: biomedical research," *Science,* February 22, 1974, pp. 715–722.

A. S. Kubo and D. J. Rose. "Disposal of nuclear wastes," *Science,* December 21, 1973, pp. 1205–1211.

Robert Gillette. "Radiation spill at Hanford: the anatomy of an accident," *Science,* August 24, 1973, pp. 728–730.

John O. Blomeke, Jere P. Nichols, and William C. McClain. "Managing radioactive wastes," *Physics Today,* August 1973, pp. 36–42.

William A. Higinbotham. "Nuclear safeguards: 2. The US program," *Physics Today,* November 1969, pp. 40–44.

J. E. Logsdon and J. W. N. Hickey. "Radioactive waste discharges to the environment from a nuclear fuel reprocessing plant," *Radiological Health Data and Reports,* June 1971, pp. 305–312.

Bernard W. Sharpe. "Nuclear safeguards: 1. The IAEA program," *Physics Today,* November 1969, pp. 33–37.

22   William R. Corliss. *Neutron Activation Analysis,* 1964. UAS

Robin P. Gardner and Ralph L. Ely, Jr. *Radioisotope measurement applications in engineering.* Reinhold, New York, 1967.

Colin Renfrew. "Carbon 14 and the prehistory of Europe," *Scientific American,* October 1971, p. 63.

W. H. Wahl and H. H. Kramer. "Neutron-activation analysis," *Scientific American,* April 1967, p. 68.

Chapter(s)   *Reference*

N. A. Matwiyoff and D. G. Ott. "Stable isotope tracers in the life sciences and medicine," *Science*, September 21, 1973, pp. 1125–1133.

22, 23   John H. Lawrence, Bernard Manowitz, and Benjamin S. Loeb. *Radioisotopes and Radiation.* McGraw-Hill, New York, 1964. Recent advances in medicine, agriculture, and industry.

22–25   Glenn T. Seaborg. *Peaceful Uses of Nuclear Energy.* USAEC Division of Technical Information Extension, Oak Ridge, Tenn., 1970.

23   Gordon Brownell and Robert J. Shalek. "Nuclear physics in medicine," *Physics Today*, August 1970, pp. 32–38.

Björn Sigurbjörnsson. "Induced mutations in plants," *Scientific American*, January 1971, p. 86.

V. T. Stannett and E. P. Stahel. "Large scale radiation-induced chemical processing," *Annual Review of Nuclear Science*, Vol. 21, 1971.

24   Lynn E. Weaver (Ed.) *Education for peaceful uses of nuclear explosives.* The University of Arizona Press, Tuscon, Ariz., 1970.

25   R. W. Bussard and R. D. DeLauer. *Fundamentals of Nuclear Flight,* McGraw-Hill, New York, 1965.

"Nuclear rockets," *Nucleonics 16*, July 7, 1958, pp. 62–75.

26   John R. Clark. "Thermal pollution and aquatic life," *Scientific American 220*, March 3, 1969, pp. 18–27.

Michel d'Orival. *Water Desalting and Nuclear Energy.* Verlag Karl ThiemigKG, Munich, 1967.

Merril Eisenbud and George Gleason (Eds.) *Electric Power and Thermal Discharges.* Gordon & Breach, Science Publishers, New York, 1969.

Engineering for Resolution of the Energy-Environment Dilemma. Committee on Power Plant Siting, National Academy of Engineering, Washington, D.C., 1972.

Nuclear Power in the South. A Report of the Southern Governors' Task Force for Nuclear Power Policy, 1970.

Riley D. Woodson. "Cooling towers," *Scientific American*, May 1971, p. 70.

Joseph Barnea. "Geothermal power," *Scientific American*, January 1972, p. 70.

Charles A. Berg, "Energy conservation through effective utilization," *Science*, July 13, 1973, pp. 128–138.

Eric S. Cheney, "U.S. Energy resources: limits and future outlook," *American Scientist*, January–February 1974, pp. 14–22.

Floyd L. Culler, Jr. and William O. Harms. "Energy from breeder reactors," *Physics Today*, May 1972, pp. 28–39.

Electric Power and the Environment. A Report Sponsored by The Energy Policy Staff, Office of Science and Technology, August 1970, Supt. of Doc., U.S. Government Printing Office, Washington, D.C.

M. King Hubbert. "The energy resources of the earth," *Scientific American*, September 1971, p. 60.

John J. Murphy (Ed.) *Energy and Public Policy—1972.* The Conference Board, New York, 1972. A report on a conference about energy supply and demand.

Chapter(s)   *Reference*

Resources and Man. National Academy of Sciences—National Research
Council, W. H. Freeman and Co., San Francisco, 1969. Especially
Chapter 8, "Energy Resources" by M. King Hubbert.

Ralph Roberts. "Energy sources and conversion techniques," *American
Scientist*, January–February 1973, pp. 66–75.

Chauncey Starr. "Energy and power," *Scientific American*, September
1971, p. 36.

Kenneth F. Weaver. "The search for tomorrow's power," *National
Geographic*, Vol. 142, No. 5, November 1972, pp. 650–681.

W. E. Winsche, K. C. Hoffman, and F. J. Salzano. "Hydrogen: its future
role in the Nation's energy economy," *Science*, June 29, 1973, pp.
1325–1332.

Index

273

TITLES IN THE PERGAMON UNIFIED ENGINEERING SERIES